Polymer Science and Technology

www.novapublishers.com

Polymer Science and Technology

A Study of Polymer Dynamics by Solid-State NMR
John Nobleman (Author)
2021. ISBN: 978-1-53619-715-0 (Hardcover)
2021. ISBN: 978-1-53619-803-4 (eBook)

What to Know about Lignin
María González Alriols, PhD
Jalel Labidi, PhD
M. Özgür Seydibeyoğlu, PhD (Editors)
2021. ISBN: 978-1-53619-152-3 (Hardcover)
2021. ISBN: 978-1-53619-222-3 (eBook)

Encyclopedia of Polymer Composites
Mikhail Lechkov
Sergej Prandzheva (Editors)
2010. ISBN: 978-1-60741-717-0 (Hardcover)
2010. ISBN: 978-1-61122-476-4 (eBook)

Polydimethylsiloxane: Structure and Applications
Philip N. Carlsen (Editor)
2020. ISBN: 978-1-53617-590-5 (Hardcover)
2020. ISBN: 978-1-53617-611-7 (eBook)

Polymer Materials with Smart Properties
Maria Bercea (Editor)
2013. ISBN: 978-1-62808-876-2 (Hardcover)
2013. ISBN: 978-1-62808-877-9 (eBook)

More information about this series can be found at
https://novapublishers.com/product-category/series/polymer-science-and-technology-series/

V. A. Ryzhov

Far Infrared and Terahertz Spectroscopy of Polymers

DOI: https://doi.org/10.52305/YKKP7083

NOTICE TO THE READER

The Publisher has taken reasonable care in the preparation of this book, but makes no expressed or implied warranty of any kind and assumes no responsibility for any errors or omissions. No liability is assumed for incidental or consequential damages in connection with or arising out of information contained in this book. The Publisher shall not be liable for any special, consequential, or exemplary damages resulting, in whole or in part, from the readers' use of, or reliance upon, this material. Any parts of this book based on government reports are so indicated and copyright is claimed for those parts to the extent applicable to compilations of such works.

Independent verification should be sought for any data, advice or recommendations contained in this book. In addition, no responsibility is assumed by the Publisher for any injury and/or damage to persons or property arising from any methods, products, instructions, ideas or otherwise contained in this publication.

This publication is designed to provide accurate and authoritative information with regard to the subject matter covered herein. It is sold with the clear understanding that the Publisher is not engaged in rendering legal or any other professional services. If legal or any other expert assistance is required, the services of a competent person should be sought. FROM A DECLARATION OF PARTICIPANTS JOINTLY ADOPTED BY A COMMITTEE OF THE AMERICAN BAR ASSOCIATION AND A COMMITTEE OF PUBLISHERS.

Additional color graphics may be available in the e-book version of this book.

Library of Congress Cataloging-in-Publication Data

ISBN: 978-1-68507-662-7

Published by Nova Science Publishers, Inc. † New York

Contents

Preface

The presented work summarizes and systematizes an extensive experimental material of the results of studying polymers using spectroscopy in the low-frequency infrared region.

Today, spectroscopic studies in the far infrared region are becoming an important tool for characterizing the physical properties of polymers, determined by molecular dynamics and the level of molecular interactions. Low-frequency spectroscopy of intermolecular interactions is the original and most informative source and criterion for establishing the presence of a hydrogen bond in biological substances, multiplets and clusters in ionomers, a criterion for crystallinity, etc.

Far IR spectroscopy has proven to be productive in deciphering the molecular nature of solid-state (δ, β, and γ) relaxation transitions in polymers. This was the result of (1) evaluating the potential barriers and sizes of molecular motion units from the spectra, (2) finding empirical correlations between the spectral parameters and molecular characteristics of polymers, (3) comparing the results with activation barriers for relaxation transitions.

List of Abbreviations

PS	polystyrene
PTFS	poly(trifluorostyrene)
PPFS	poly(pentafluorostyrene)
PMS	poly(α-methylstyrene)
PP	polypropylene
PTFE	poly(tetrafluoroethylene)
POE	poly(oxyethylene)
POM	poly(oxymethylene)
PVA	poly(vinyl acetate)
PAN	polyacrylonitrile
PCS	polychlorostyrene
PVF	poly(vinyl fluoride)
PVF2	poly(vinylidene fluoride)
PBMA	poly(n-butyl methacrylate)
POMA	poly(octyl methacrylate)
PDMA	poly(decyl methacrylate)
PCHMA	poly(cyclohexyl methacrylate)
PC	polycarbonate
PE	polyethylene
PMMA	poly(methyl methacrylate)
PVC	poly(vinyl chloride)

List of Symbols

$\nu_1, \nu_2, \ldots, \nu_{12}$, $\nu_\alpha, \nu_s, \nu_\delta, \nu_\gamma, \delta_\alpha$, $\delta_s, \delta_\omega, \delta_r, \gamma_t$	Nomenclature of normal vibrations
ν_{sk}	Wave number, cm^{-1} for skeletal vibrations
ν_{loc}	Wave number, cm^{-1} for local torsional vibrations
ν_{libr}	Wave number, cm^{-1} for librational vibrations
A	Total intensity of the absorption
A_{sk}	Intensity of the skeletal vibration band
$\alpha(\nu)$ or $k(\nu)$	Absorption coefficient
$\alpha_{sk}(\nu)$	Absorption coefficient of the skeletal vibration band
ν	Wave number in cm^{-1}
$\bar{\nu}$	Frequency in Hz
ω	Angular frequency in Hz
c	Light velocity
k	Boltzmann constant
k_1, k_2, k_s, k_b, k_t	Force constants of vibrations
f	Force constant in Eq. (1)
θ	Phase difference
φ	Angle of libration
ϕ	Angle of torsion vibration in backbone chain
m	Atomic mass
M	Repeated unit mass
μ_{eff}	Effective dipole moment
Req	Equivalent radius of the librational unit
I	Moment of inertia
$C^{n\div}$	Cation with charge n +
n	Number of monomer units in polymer chains
N	Number of monomer units in oligomers

S	Statistical (Kuhn) segment size of macromolecule in a number of monomer units
tan 6	Dielectric loss tangent
ε	Unelastic deformation
$\varepsilon''(\nu)$	Dielectric loss
$\alpha, \beta, \gamma, \delta$	Relaxation processes (transitions)
$T_g, T_\beta, T_\gamma, T_\delta$	Temperatures of α, β, γ and δ relaxation processes
T*	A temperature of onset (on heating) or stopping (on cooling) of conformational rearrangements in chain
T_2	Characteristic temperature in Gibbs-DiMarzio thermodynamic glass transition theory

Introduction

The subject of the book is the infrared (IR) absorption spectra of polymers in the frequency range between radio and IR, which spectroscopists usually call the far infrared (FIR) region. In the spectrum of electromagnetic radiation, the FIR region occupies a frequency range from 12 THz (400 cm^{-1}) to 0.1 THz (3.33 cm^{-1}). In the past few decades, due to the success of terahertz technologies, the name of the terahertz region has been stuck in the 3 - 0.1 THz subrange.

As is known, the low-frequency region of the IR spectrum of a polymer is expected to be complex, including both intra- and intermolecular modes. The latter will be external vibrations of the chain of both rotational and translational types, while the former include both local torsional vibrations of functional groups and modes that will be delocalized along the chain. In addition, absorption in this region is determined by such phenomena as disorder-induced absorption, relaxation and non-resonance processes in the polymer. Absorption due to the presence of both accidental and deliberately introduced impurities also plays a significant role.

The difficulty in interpreting the FIR spectrum of a polymer is one of the reasons for the lag in low-frequency IR spectroscopy. Another is that for a long period there was no suitable experimental technique, and only now, in connection with the advent of serial IR-Fourier spectrometers (FT-IR) and terahertz spectrometers (THz-TDS) (see Application 3), the situation changes somewhat.

The spectroscopy of polymers in the low-frequency infrared region was held back for a long time also because polymer spectroscopists, in the absence of a well-developed theory of the FIR spectra of polymers and sufficient experimental material, were forced to use the concept of group frequencies. However, in the low-frequency IR region, this simple concept needs to be refined, taking into account intra- and intermolecular bonding, leading to collective excitations similar to phonons in crystals.

As a result, most of the studies on the FIR spectra of polymers published in the last few decades are largely limited so far only to illustrate the possibilities of using this spectral technique.

Attempts to bring together the available data on the FIR spectra of polymers were previously carried out in reviews [1, 2, 3], which include only data limited to the 70s of the last century. A small review (4) of 1983 clearly does not claim to be complete and contains only 30 references demonstrating the basic principles of long-wave IR spectroscopy of polymers. The last most comprehensive review of the work on FIR spectra of polymers over the previous 10-20 years was published in Advances in Polymer Science, vol. 114 in 1994 [136].

In this book, which supplements review [136], for the first time, we collect practically all works published in the last 10–20 years on the spectra of polymers in the low-frequency IR region, including the terahertz range. Particular attention is paid to obtaining information on torsional-vibrational dynamics and intermolecular interactions of various nature, direct estimates of the conformational state of chains and molecular mechanisms of relaxation transitions in polymers.

Chapter 1

Features of Infrared Spectra of Polymers

1.1. A Theoretical Approach to the Interpretation of Spectra

The discussion of FIR spectra of polymers should begin with their characteristic feature due to the structural and morphological specificity of polymers. It consists in the fact that polymer molecules are chains of repetitive chemical groups (monomer units). In the simplest case, such a chain is linear, and although macromolecules can also branch and form three-dimensional networks, in general it is a linear system with strong covalent bonds along the chain and weak ones between chains. The composition and degree of polymerization affect the conformational structure of macromolecules, formed by rotation around chemical bonds, which in turn determines the packing of chains in a condensed state. The latter will be amorphous or crystalline, but most often partially crystalline, usually represented by a two-phase model with crystallites embedded in an amorphous matrix.

This specificity of polymers affects their spectra and requires the use of special methods for interpretation. The most important of these is group theory, which allows one to analyze with a good approximation the vibrational modes of macromolecules in terms of local symmetry. In this concept of normal vibrations of atoms, the motion is localized within the repeating link of the macromolecule and is controlled by the structure of the latter: the masses involved in the vibration and the corresponding force constants. The influence of other chemical bonds is assumed to be negligible or at least easily calculated [5, 6]. In the mid-IR region, methods have been developed for calculating the curves of the spectral distribution of the absorption coefficient for a number of polymers, which provide an accuracy that is quite satisfactory for practical purposes [7].

However, when considering lower-energy motions in macromolecules, this approach is insufficient. It becomes necessary to take into account not only intramolecular, but also intermolecular interactions, which cause the appearance of collective excitations in the polymer, similar to phonons in crystals.

For the description of low-frequency vibrations, an approach that comes not from the spectroscopy of individual molecules, but from the spectroscopy of crystals is more justified in most cases. Let us illustrate what has been said by an example of calculating skeletal vibrations in a macromolecule.

In the approach coming from crystal spectroscopy [5, 8, 9], a macromolecule is considered as a chain of oscillating point masses *M* separated by a distance d and linked by chemical bonds: the latter determine the stretching, deformation, or torsional force constant of these oscillations f.

As a result, the problem is reduced to the calculation of a linear monatomic chain, its normal vibrations. The oscillation frequency of such a chain is a periodic function of the wave number *k*:

$$\nu = (1/\pi) \bullet (f/M)^{1/2} \sin (\pi kd) \quad (1)$$

The plot of frequency *versus* wave vector is known as a dispersion curve. The first period of this function represents the 1st Brillouin zone. For a linear diatomic chain, in which all bonds are identical, and the atoms are located at an equal distance, but have different masses M_1 and M_2:

$$\nu^2 = (f/4\pi^2) \{(1/M_1 + 1/M_2) \pm [(1/M_1 + 1/M_2)^2 - 4/M_1M2 \sin^2 (\pi kd)]^{1/2}\} \quad (2)$$

and there are two dispersion curves, of which one passing through the origin is the acoustic branch (its frequencies are related to the region of sound and ultrasonic frequencies), and the other is the optical branch with the frequencies of the IR range. Modes of the optical branch: corresponding to $k = 0$, characterize the stretching of the bond between the atoms of the chain. Acoustic vibrations at $k = 0$ do not contribute to the change in the bond dipole moment and, therefore, are inactive.

When a repeating unit contains *n* atoms: there are 3*n* separate dispersion curves (branches) of normal vibration modes. Three of them correspond to acoustic modes, and the rest 3*n*-3 to optical ones. The latter can be, as a first approximation, subdivided into intramolecular, characterizing a repeating unit, and skeletal or, in the terminology of solid-state physics, lattice vibrations.

1.2. Dispersion Ratios for Polyethylene

As the simplest example, consider an endless PE chain, in which the repeating unit CH_2 - CH_2 contains 6 atoms and therefore is characterized by 18 normal vibrations. Figure shows 9 corresponding dispersion branches. Branches v_1 and v_6 correspond to symmetric and antisymmetric stretching vibrations in the CH_2 group, v_2 - its bending vibrations. Branch v_3 characterizes fan vibrations, v_4 - stretching of the C-C bond, v_5 - deformation skeletal vibrations; v_7 and v_8 are joint torsional and rocking vibrations of the CH_2 group and, finally, v_9 are skeletal torsional vibrations. The intersection of these branches with the ordinate at $k = 0$ (phase difference $\theta = 0$) and the ordinate $k =$ (phase difference $\theta = \pi$) gives the frequencies active in IR and Raman spectra.

The type of branches reflects the sensitivity of this type of oscillation to the phase difference between the oscillators in the chain and the dependence of the modes of these oscillations on the chain length. As can be seen from Figure 1, the most complete correspondence between the calculated and observed the frequencies should be expected for branches of the v_1, v_2, and v_6 types, setting highly localized fluctuations. Skeletal vibrations (branches v_4 v_5 and v_9) are less characteristic, and the frequencies v_5 and v_9 at $\theta = 0$ and π generally turn out to be zero, requiring, to identify these vibrations in the spectra, the calculation of frequencies potentially active for a chain of finite dimensions.

This calculation is difficult, because when monomeric units are repeated in a chain in a regular manner and are energetically similar, the possibility of a resonant unification of the oscillatory motion arises. This leads to the appearance of a series of vibrational modes characterizing the length of the monomer units of the chain united by interactions.

Constraints imposed by the selection rules, a sharp drop in the intensity of modes far from the edge of the dispersion curve; as well as the defectiveness of real chains and the resolution of the device, however, leads to the fact that in the spectrum of such a series of bands there will correspond one asymmetric absorption band characterizing a section of the chain of finite length [10, 11] or 'vibrational segment', according to terminology [12].

Comparison of the calculated frequencies of skeletal vibrations with the experimentally observed frequencies shows their satisfactory agreement in the PE spectrum in the mid-IR region [5]. Thus, the frequency branch for optically active skeletal vibrations v_4 intersects with the ordinate at $v_0 = 1065$

cm^{-1} and ν_π = 1135 cm^{-1}. These calculated values correspond to the bands at 1061 cm^{-1} and 1131 cm^{-1} in the Raman spectrum of the PE.

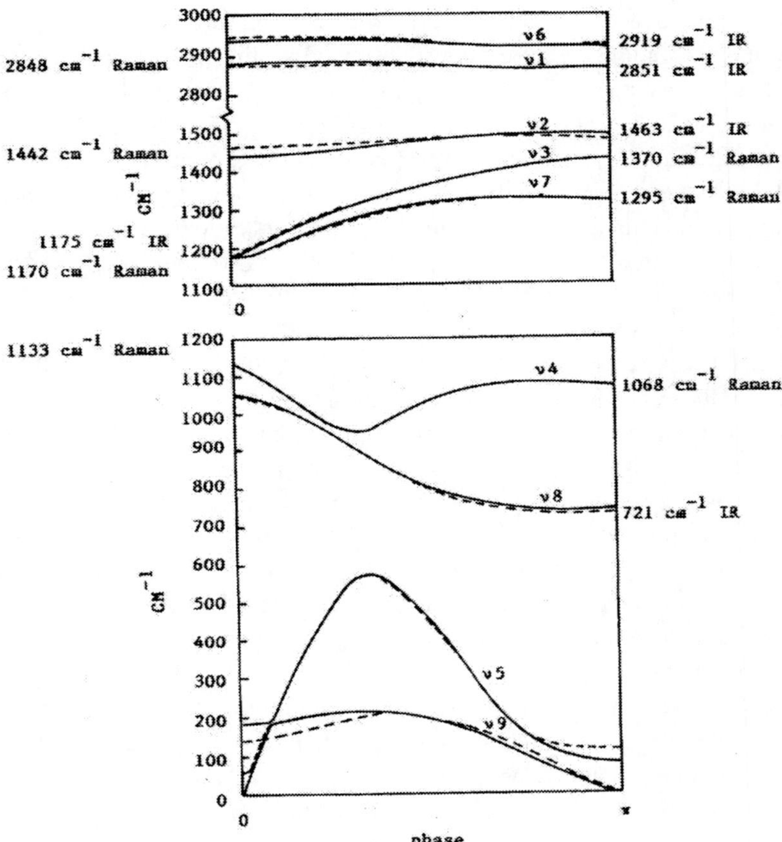

Figure 1. Dispersion curves and vibrational spectra for an infinite polymethylene chain (*solid lines*) and an orthorhombic PE crystal. (*Dashed lines* show oscillations perpendicular to the *x*-axis [6]).

With an increase in the identity period (up to four CH_2 groups), which is equivalent to taking into account the interaction between neighboring units, frequencies appear: corresponding to bending vibrations at $\nu_{+\pi/2}$ = 983 cm^{-1} and $\nu_{-\pi/2}$ = 430 cm^{-1}. In addition, in this case, there will be out-of-plane skeletal oscillations with a frequency

$$\nu^2_{\pi/2} = k_t\,(4 - 2\cos\theta - 4\cos 2\theta + 2\cos 3\theta)/4\pi^2 m, \quad (3)$$

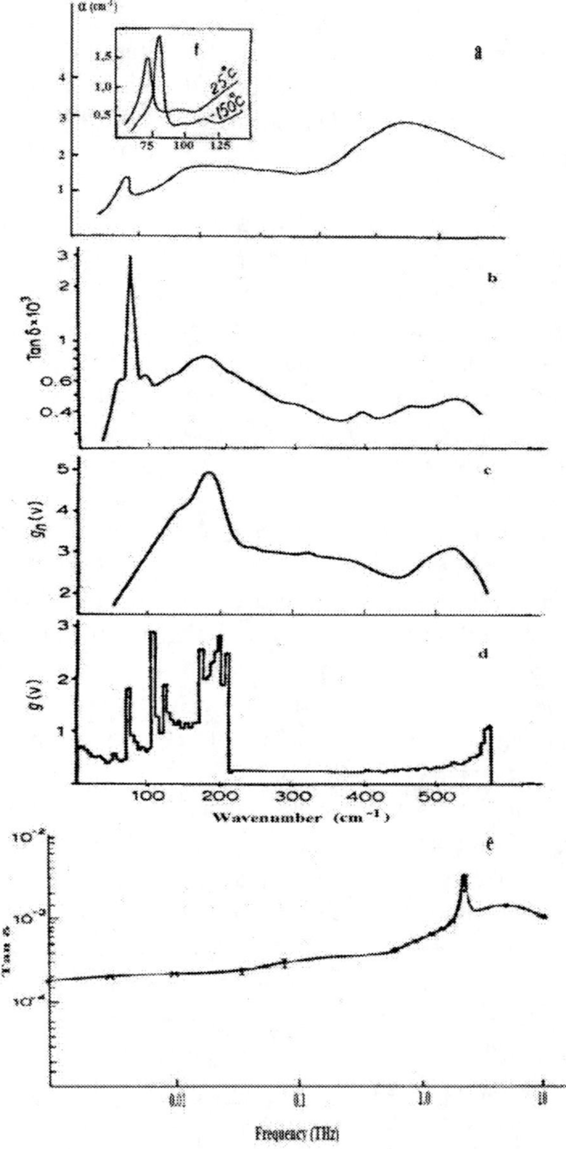

Figure 2. a-f. Low-frequency IR spectra of PE. 2a [2], 2f [17], 2b [19] neutron scattering spectrum, 2c [18] dielectric loss tangent, 2d [18] calculated spectrum of vibrational density states, 2e is the dielectric loss tangent according to [19].

where m is the mass of CH_2 group and k_t is the force constant of rotational motion around of the C – C bond. Assuming the latter to be equal, according to [5], k_t = 20N/m, we obtain the calculated frequency of skeletal torsional vibrations $\nu_{\pi/2}$ = 140 cm^{-1} in PE. In the FIR spectrum of PE, this vibration apparently corresponds to a broad absorption band centered at 200 cm^{-1} (Figure 2a).

When passing from the spectrum of an individual chain to the spectrum of a crystal, taking into account vibrations perpendicular to the chain leads to the separation of dispersion branches, which is shown in Figure 1. Solid lines in this case will correspond to the branches of vibrations along the chain (axis α of the crystal), and dashed lines to branches of vibrations perpendicular to it (axis β of the crystal). The intersection of the branches ν_5 and ν_9 with the ordinate ($\theta = 0$ and $\theta = \pi$) gives in this case 8 frequencies, of which 3 are acoustic with zero frequency, and the remaining 5 are lattice modes. Of these, the mode $\nu^a{}_5$ ($\theta = \pi$) = 73 cm^{-1} is active in IR absorption, first of all, which corresponds to the translational vibration of the crystal lattice of the PE, which is actually observed in the FIR spectrum of the PE. This assignment is consistent with the dichroism of this band [13] and its shift in the spectrum of deuterated PE [14]. As the temperature rises from 12 to 400 K, the band shifts in frequency from 81 to 68 cm^{-1} due to the expansion of the crystal [15], and with increasing pressure in the opposite direction. This result allows one to judge the lattice anharmonicity and the Gruneisen parameter of the PE [16].

The second mode active in the FIR region, $\nu^b{}_5$ ($\theta = \pi$) = 108 cm^{-1}, is observed in the PE spectrum at low temperatures (Figure 2f, inset). This band is low-intensity and insufficiently studied.

Comparing the data of these calculations with the spectrum of partially crystalline PE in the wavelength range of 50 - 500 cm^{-1} (Figure 2a), it should be recognized that individual calculated optical modes do not reflect the entire experimentally observed picture. In reality, the spectrum is closer to the spectrum of the density of vibrational states (Figure 2d), obtained taking into account nonzero phase shifts and “quasi-lattice” modes characterizing short-range order in the amorphous regions of the polymer. Figure 2b shows the spectrum of neutron scattering in PE, in which all modes are allowed and, therefore, it better coincides with the spectrum of the density of vibrational states [18]. In Figure 2c and 2e show the low-frequency spectrum of a PE in a wide frequency range, presented as a plot of the dependence of the tangent of the dielectric loss angle (tan δ) versus the wavenumber (in cm^{-1}) or frequency (in Hz) [18, 19]. In this loss spectrum, in addition to the lattice absorption at 73 cm^{-1} and absorption at 200 cm^{-1} discussed above, one can see two broad loss regions in the form of a background at 10^8 - 10^{10} and 10^{12} - 10^{13} Hz, which will be discussed in section 5.1.

1.3. Other Highly Crystalline Polymers

The computational approach considered for PE is also used when assigning absorption in the FIR spectra of other highly crystalline polymers, such as PP [2.20], PTFE [21, 24, 26], POE and POM [22].

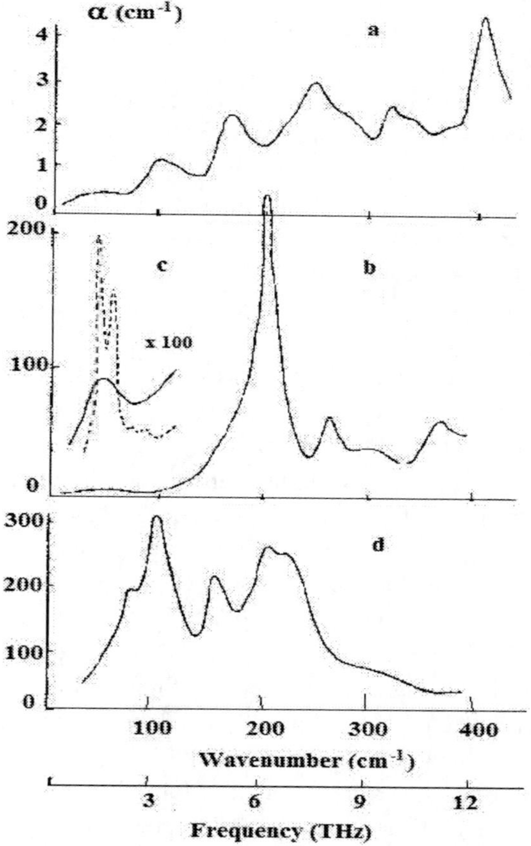

Figure 3. FIR spectra of highly crystalline polymers. Isotactic PP (a) [20], PTFE (b and c) [24.62] and POM (d) [152].

According to calculations performed for isotactic and syndiotactic PP [20] in the low-frequency region, one can expect the appearance of modes characterizing bending vibrations of the skeleton and internal rotation in the main chain around the C - C bonds. Since these vibrations are not localized within one monomer unit, the corresponding absorption bands are most satisfactorily identified in the PP spectrum precisely at availability is enough high concentration of the crystalline phase. The number of experimentally observed absorption bands here is much larger than that predicted by the model of an isolated macromolecule, and it is assumed that they correspond mainly to mixed fluctuations. Thus, observed in the spectrum of isotactic PP (Figure 3a) the band with a maximum at 398 cm^{-1}, according to [20, 21], corresponds to the calculated frequency 392 cm^{-1}, which characterizes the bending vibration of C-CH-C groups, 30% symmetric (δs) and 40% antisymmetric (δa). The 321 cm^{-1} band (calculated at 314 cm^{-1}) is (δa) by 60%, and the 251 cm^{-1} band (calculated at 272 cm^{-1}) is 40% (δs) and 20% (δa). The low-intensity band at 210 cm^{-1} (calculated at 207 cm^{-1}) characterizes by 95% the torsional vibrations of the CH_3 group, and the band at 169 cm^{-1} (calculated at 162 cm^{-1}) by 40% corresponds to the asymmetric and 30% symmetric vibration of the C- CH-C group and 15% torsional vibration around the C- C bond (τ_{cc}).

Finally, the 106 cm^{-1} band (90 cm^{-1} calculation) characterizes 70% torsional skeletal vibrations.

It is not possible to explain on the basis of calculations for an isolated macromolecule only the broad band 55 cm^{-1} (48 cm^{-1} at 100 K), presumably attributed in [20] to the lattice vibration. It can be expected that this absorption characterizes the amorphous part of the polymer, similar to the background absorption in PE in the range 10–100 cm^{-1} [23]. This so-called "disorder-induced" quasi-lattice absorption is now found to be inherent in the FIR spectra of all disordered condensed media will be discussed below.

The computational approach turned out to be fruitful in the case of one more highly crystalline PTFE polymer. PTFE can be described as PE in which all of the hydrogens are replaced by fluorine atoms, but the solid phase of PTFE is different from that of PE. Here macromolecules do not form flat zigzag chains, but are twisted into a spiral. There is also a difference in intermolecular forces. A discussion of the absorption spectrum of such a complicated helical macromolecule in terms of dispersion relations was done in [24]. The main result is as follows:

The band at 385 cm^{-1} (380 cm^{-1} calculated) is 70% deformation (ν_δ) vibration of the CF_2 group and 25% wagging mode (γ_w); the 308 cm^{-1} band

(300 cm^{-1} calculation) is mainly its rocking vibration (γ_r); band at 277 cm^{-1} (calculated 280 cm^{-1}) 60% rocking vibration (γ_r) of the CF_2 group and 20% stretching (ν_5) skeletal vibration of the C - C bond. The most intense band in the FIR spectrum of PTFE, centered at 203 cm^{-1} (calculation 192 cm^{-1}), refers to purely torsion vibrations of the CF_2 group (γ_t mode). The branch corresponding to the torsional vibrations of the chain skeleton (mode ν_9), according to the calculation, should be represented by single band in the spectrum. However, in the FIR spectrum of PTFE with a high degree of crystallinity, measured at low temperatures, a number of narrow bands appear at 46, 55, 70, and 85 cm^{-1} (Figure 3c), merging at a temperature of T $\geq$ 200 C, i.e., above the temperature of the solid-phase transition, into a wide band with a maximum near 50 cm^{-1}. According to [25], the latter can be attributed, by analogy with the assignment of such absorption in the FIR spectra of PE and PP, to liquid-lattice (quasi-lattice) absorption. The band splits into a number of bands with decreasing temperature, corresponding to the transition to true lattice vibrations in the crystalline regions of the polymer.

The FIR spectrum of PTFE is an excellent example of the capabilities of FIR spectroscopy for plotting dispersion curves from experimental data. The basis for this procedure is a violation of the selection rules, and not only in the absence of crystalline order, but also when the chain becomes so short that the boundary conditions cease to operate. As an example, we refer to [26] on the study of low-frequency IR spectra of PTFE oligomers.

The theoretical approach also makes it possible to assign bands in the FIR spectrum of highly crystalline POE (Figure 3d), where the band at 215 cm^{-1} is referred to skeletal bending vibrations [22] (calculated at 211 cm^{-1}). The 165 cm^{-1} band (calculated at 162 cm^{-1}) is associated with torsional vibrations around the C-C bond, while the 106 cm^{-1} band (calculated at 108 cm^{-1}) is mainly a characteristic of torsional skeletal vibrations around the C-O bond.

Similarly, experiments and calculations for POM (22) made it possible to assign the 428 cm^{-1} band (calculated at 444 cm^{-1}) to the deformation of the C-O-C angle (calculated at 444 cm^{-1}), and to the torsional skeletal vibrations of the 304 cm^{-1} band (calculated at 253 cm^{-1}), at 130 cm^{-1} (calculation 135 cm^{-1}) and at 90 cm^{-1} (calculation 86 cm^{-1}). The performed assignment made it possible to reveal the important role of the conformational chain states, since skeletal modes are strongly interconnected and any change in conformation affects their frequencies.

Chapter 2

Skeletal Torsion-Deformation Vibrations in Amorphous Polymers: Theoretical Predictions and Experiment

2.1. Polymethyl Methacrylate (PMMA), Polyvinyl Chloride (PVC), and Polystyrene (PS)

Normal vibrational analysis, based on the correct structure and the correct potential field, allows one to obtain a good agreement between the modes predicted by the calculation and the observed absorption in the FIR spectra of highly crystalline polymers, despite the existence of disordered regions in them. Although, of course, the size and defectiveness of polymer crystals, imposing boundary conditions, will determine both the contour and the position of the bands in the spectrum. In addition, due to violation of the selection rules, this should also lead to broadening of the bands and the appearance of new absorption, which is not predicted by the calculation. The latter, in fact, is reflected in the FIR spectrum at frequencies corresponding to the maxima in the distribution function over vibrational states. This is the origin of the wide absorption region centered at 200 cm^{-1} in the FIR spectrum of the PE.

The successful use of theoretical methods for assigning bands in the FIR spectra of highly crystalline polymers gives grounds for its use in the case of semicrystalline and amorphous polymers. Such a calculation, for example, was performed in [27] for syndiotactic PMMA. It was carried out for a zigzag chain consisting of two alternating masses M_1 and M_2, where M_1 is the mass of the CH_2 group, and M_2 is the mass of the CH3-C-COOH3 group. Due to the alternative nature a syndiotactic chain, a repeating cell contains 4 such "masses" and, therefore, has $3N = 12$ skeletal normal vibrations, of which 4 are nondegenerate (3 translational and one torsional) and 8 degenerate (6 of them are active in IR and Raman).

The frequencies of in-plane skeletal vibrations were found by solving the equation proposed by Zbinden (8), which in this case, when only vibrations with a phase difference $\theta = 0$ and π are active, is reduced to the form:

$$\begin{bmatrix} (M_2w^2 - G) & G & 0 & 0 \\ G & (M_1w^2 - G) & 0 & 0 \\ 0 & 0 & (M_2w^2 - R - 4Q) & (4Q + R) \\ 0 & 0 & (4Q + R) & (M_1w^2 - R - 4Q) \end{bmatrix} = 0 \qquad (4)$$

and for the phase difference $\theta = \pi/2$ and $3\pi/2$ to the form:

$$\begin{bmatrix} (M_2w^2 - G - 2F) & 0 & 0 & i(N - 2P) \\ 0 & (M_1w^2 - G - 2F) & i(2P - N) & 0 \\ 0 & i(N - 2P) & (M_2w^2 - R - 2Q) & 0 \\ i(2P - N) & 0 & 0 & (M_1w^2 - R - 2Q) \end{bmatrix} = 0 \qquad (5)$$

Here $\omega = 2\pi c\nu$ is the angular frequency, and the coefficients are: $G = 2k_s Cos^2\alpha$, $R = 2k_sSin^2\alpha$, $Q = 2k_b Cos^2\alpha$, $F = 2k_bSin^2\alpha$, $P = 2k_b Cos\alpha Sin\alpha$, $N = 2k_s Cos\alpha Sin\alpha$, where α is the angle between C – C bond and the direction of the chain, and k_s is the force constant of stretching of the C-C bond and k_b is the force constant of the deformation of the C-C-C angle. The solution of the above equations gave the following frequencies $\nu_1 = 0$, $\nu_1 = 803$ cm^{-1}, $\nu_3 = 0$, $\nu_4 = 796$ cm^{-1}, $\nu_5 = 796$, $\nu_6 = 370$ cm^{-1}, $\nu_7 = 591$ cm^{-1} and $\nu_8 = 201$ cm^{-1}.

The frequencies outside the plane skeletal vibrations according to Zbinden were obtained from the equation:

$$\omega^2 = \frac{4k_t \sin^2\theta}{M_1M_2}[M_1 + M_2 \pm (M_1^2 + M_2^2 + 2M_1M_2\cos 2\theta)^{1/2}] \qquad (6)$$

where k_t is the force constant of torsional vibrations for the C-C-C-C chain. For $\theta = 0$ and π zero oscillations will be potentially active. However, for $\theta = \pi/2$ and $3\pi/2$: $\omega^2_+ = 8\ k_t/M_2$ and $\omega^2_- = 8\ k_t/M_1$ and thus the following frequencies are obtained for out-of-plane frequencies: $\nu_9 = 0$, $\nu_{10} = 182$ cm^{-1}, $\nu_{11} = 0$ and $\nu_{12} = 73$ cm^{-1}.

The calculated skeletal vibration frequencies were compared with those experimentally observed for syndiotactic PMMA. These results are summarized in Table 1. Here, for comparison, the data for PE are given.

The FIR spectrum of PMMA is shown in Figure 33, curve 1. It can be seen that the discrepancy between the calculation and experiment mainly concerns the lowest-frequency oscillations and this is attributed to a fault of the approach based on the calculation of skeletal oscillations for PE, uncertainty in the choice of force constants, and also not taking into account interchain interactions.

Table 1. Comparison of the calculated observed frequencies (cm^{-1}) of skeletal vibrations and assignment of absorption bands in the FIR spectra of PE, c-PVC and c-PMMA

Polyethycne[a] calculated	s-PVC[b]		s-PMMA[c]		Assignment
	Observed	Calculated	Observed	Calculated	
1135	1125	1054	807	803	ν (C–C) ∥ or $[\nu_+(\pi)]$[d]
1065	1096	1076	749	761	ν (C–C) ⊥ or $[\nu_+ (0)]$
983	963	993	786	796	ν (C–C) $[\nu_+(\pi/2)]$ and $[\nu_+(3\pi/2)]$
430	487	476	320	370	δ (C–C–C) in-plane $[\nu_-(\pi/2)]$ and $[\nu\ (3\pi/2)]$
140	182	–	225	182	δ (C–C–C) out-of-plane or $\nu(\pi/2)$ and $\nu(3\pi/2)$

a Ref. [15], b Ref. [28], x Ref. [27], d notation Ref. [5].

The authors of the calculation of skeletal vibrations in the FIR spectrum of PVC come to similar conclusions [5, 28]. The FIR spectrum of PVC is shown in Figure 4. In addition to the frequencies given in Table 1, bands at frequencies of 363 cm^{-1} are found in the spectrum of this polymer (calculation 349 cm^{-1} for the δ (C-Cl) mode) and at 340 cm^{-1} (calculation of 347 cm^{-1} for the γ (C-Cl) mode), as well as bands with maxima at 86 cm^{-1} and 64 cm^{-1} not following from the calculation (their assignment is condemned below).

In the FIR spectrum of PS (Figure 4), on the basis of a similar calculated assignment [5, 29], the absorption at 410 cm^{-1} (calculated at 421 cm^{-1}) was assigned to the skeletal vibration bands with the predominant contribution of ν_{9b} from the benzene ring. Weak absorption at 325 cm^{-1} (calculation 358 cm^{-1}) is related to the deformation of the C-C-C angle in the main chain. A more intense band with a complex structure with a maximum at 216 cm^{-1} is attributed to intra-molecular deformation vibration of the benzene ring (calculation 243 cm^{-1}), as well as the deformation vibration of the backbone skeleton. Absorbance at 245 cm^{-1} and 80 cm^{-1} is not interpreted within this approach. Its assignment is discussed below.

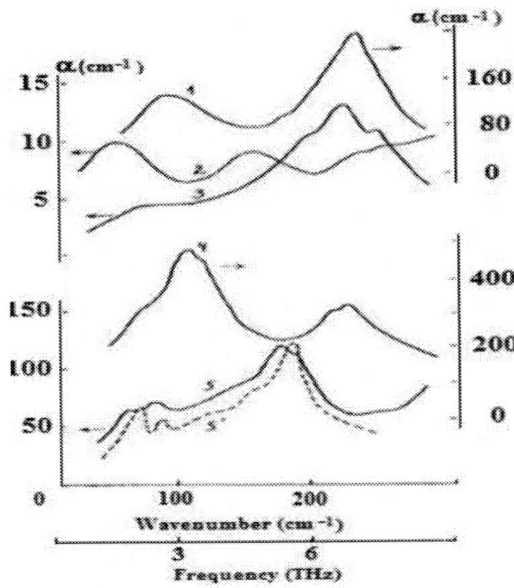

Figure 4. FIR spectra of polymers (1) PVA [40], (2) PCS [31], (3) PS [31], (4) PAN [(31], PVC at T = 293K (5) and T = 78K (5`) [33].

2.2. Correlation Approach to Interpretation of Skeletal Vibrations in the Spectra of Amorphous Polymers

Other attempts at a computational approach to the assignment of absorption bands in the FIR spectra of more complex polymers indicate the insufficient efficiency of this approach to the interpretation of low-frequency skeletal vibrations [5, 7, 8, and 21].

So, at one time, when working in the mid-IR region of the spectrum, problems arose that required a more complete interpretation of the spectra than the group-frequency approximation allows. However, the use of in-depth theoretical analysis of non-characteristic oscillations for solving real problems turned out to be limited in this situation. The way out was to find the 'spectrum-structure-property' relationship experimentally, with the help of empirical correlations found during research.

In the case of FIR spectra, where the theory has been developed even less, the experimental correlation approach, as it seems at present, is the only effective one for solving a number of specific problems. The possibilities of this approach are discussed here by the example of studying glassy polymers from their FIR spectra [30.31].

The objects of study included a large group of linear and dosed-crosslinked polymers, copolymers, and oligomers. In addition to the aforementioned PS, PMMA, and PVC, the spectra of poly (chlorostyrene) (PCS), poly (butyl methacrylate) (PBMA), poly (cyclohekyl methacrylate) (PCGMA), poly (vinyl fluoride) (PVF), poly-α-methyl styrene (PMS), polyacrylonitrile (PAN), polycarbonate (PC), styrene-methacrylic acid copolymers (SMAC), crosslinked copolymers of styrene with divinylbenzene (SDV) and methyl methacrylate with dimethyl ethylene glycol (MMA-DMEG), as well as a number of oligomethacrylates, oligocarbonates, and oligo-α-methylstyrenes. The conditions for preparing samples and recording spectra are given in [30–33].

Bands of skeletal torsion-deformation vibrations in the obtained spectra could be expected in the range of 150-350 cm^{-1}. It was also assumed that in some cases they can characterize mixed vibrations in the chain with the participation of internal modes of side groups, and the spectral parameters of the absorption bands will depend on the conformational structure of macromolecules. The analysis and assignment of bands in the obtained spectra were carried out taking into account the calculated and experimental literature data.

In the FIR spectrum of PS, a band at 216 cm^{-1} was previously observed, which was attributed to the bending vibrations of the benzene ring [5] or to the joint vibration of the ring with the bending vibration of the main chain [7, 34]. As can be seen from Figure 4, the complex contour of this band includes at least two components at 216 and 245 cm^{-1}. The corresponding pair of bands shifted to high frequencies (230 ± 5 cm^{-1} and 260 ± 5 cm^{-1}) was observed in [29] for crystalline models of PS (di- and trimer). Calculations made it possible to classify the lowest frequency a doublet band to deformation of the benzene ring, and a higher frequency band to torsional skeletal vibration in the main chain. The peak at 250 cm^{-1} characterizing the motion of the PS main chain is found in the PS neutron scattering spectrum [35].

The absorption band at ca 150 cm^{-1} and 328 cm^{-1} in the PCS spectrum (Figure 4, curve 2) is due to the out-of-plane vibration of the chlorobenzene rings, according to the assignment in the Raman spectra of polychlorostyrenes [36]. Absorption in the range 240-265 cm^{-1} can also be related, by analogy with PS, to skeletal torsional vibrations in the chain. The same assignment as in the spectrum of the PS can be made for the bands at

226 and 245 cm^- of the PMS (Figure 12a). Absorbance in PS, PMS and PCS below 100 cm^{-1} is discussed below.

Figure 12b shows the spectra of PC and its oligomers. By analogy with the spectra of the above-considered polymers with benzene rings, the bands at frequencies of 220 and 260 cm^{-1} are obviously associated with torsional skeletal vibrations in the chains. Of course, such an assignment is not unambiguous, but it is quite probable and more informative than the statement that absorption in the FIR spectrum of PC in the region of 25–400 cm^{-1} is due to multiphoton processes in this polymer [37].

Above, we discussed the results of calculating the frequencies of low-frequency oscillations in PMMA [27]. A band at 225 cm^{-1} in its FIR spectrum is associated with skeletal torsion-deformation vibrations (Figure 34, curve 1). The doublet structure of this band is due to the existence of two rotational isomeric states of the main chain regions [38]. The FIR spectra of other polymethacrylates (PBMA, PCGMA, etc.) are close to the FIR spectrum of PMMA. The skeletal vibration bands are shifted to 210 and 208 cm^{-1}, respectively. A similar absorption band of skeletal vibrations lies at 232 cm^{-1} in the PVA spectrum (Figure 4, curve 1). This band is structure less, since there is no isomerism in PVA.

In the FIR spectrum of PAN (Figure 4, curve 4), the skeletal torsional vibrations in the chain include a conformationally sensitive doublet band with maxima at 240 and 257 cm^{-1}, which in [9, 41] also refers to deformation of C - C - N and C-C-CN groups.

The lowest frequencies of skeletal vibrations are observed in the spectra of PVC and PVF.

In the PVC spectrum: the bands of skeletal torsional vibrations are located at 185 and 192 cm^{-1} and at 165 cm^{-1} in PVF, which, generally speaking, is not far from the calculated value of the frequency of these vibrations 140 cm^{-1} in PE [5], which indicates a low degree of perturbation of the main chain caused by fluorine and chlorine atoms. However, experimentally in the FIR spectrum of PE, the indicated band centered at 200 cm^{-1} is very low in intensity, which means that the IR activity of torsional skeletal vibrations is significant only in the presence of polar side groups. Some discrepancy between the frequency of skeletal vibrations predicted by normal coordinate analysis and the position of the 185 cm^{-1} band in the PVC spectrum is discussed based on the study of the FIR spectra of PVC with different degrees of tacticity in [71].

2.3. Connection of Skeletal Torsional Vibrations with Intermolecular Interactions in Amorphous Polymers

The presented experimental data indicate that the position of the skeletal vibration bands in the region of 150 - 300 cm^{-1} depends on the chemical structure of the studied carbo-chain polymers. It was interesting to reveal the relationship between the spectral parameters of these bands and the molecular characteristics of polymers, such as constant dipole moment (μ_{eff}), mass of a repeating unit (M), and cohesion energy, a parameter characterizing the level of intermolecular interactions in a polymer (E_{coh}).

As a result, an unambiguous relationship was found between the position of the maximum of the bands of low-frequency skeletal vibrations and E_{coh} (Figure 5): $\nu_{max} \sim (E_{coh})^{1/2}$.

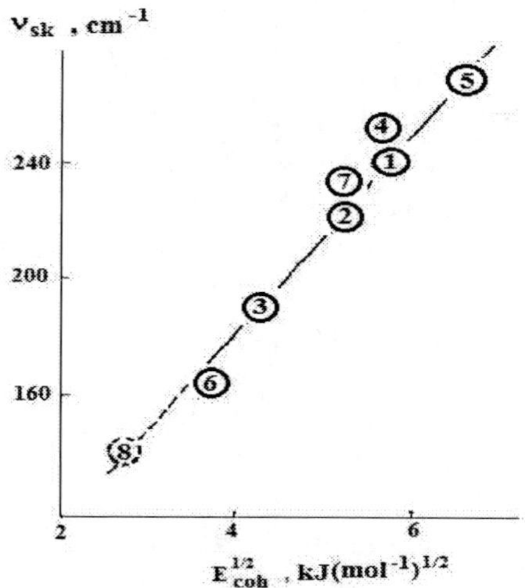

Figure 5. Dependence of the frequency of skeletal vibrations on the cohesion energy of polymers (1) PS, (2) PMMA, (3) PVC, (4) PAN, (5) PCS, (6) PVF, (7) PVA, (8) PE calculation [5].

This proportionality directly shows that the force constant of skeletal vibrations of the main chain depends on the level of molecular interactions in the polymer. A change in the energy of these interactions, for one reason or another, should also change the parameters of the considered absorption, which is indeed observed in the FIR spectra of pre-deformed polymers. Actually, an inelastic strain of ε = 15-50%, leading to an increase in enthalpy in them [42] and molecular motion (intensity of mechanical relaxation) [43, 46], simultaneously leads to an increase in the half-width and intensity of the bands of torsional skeletal vibrations [44]. The latter was caused, often, by an increase in amplitudes (anharmonicity) vibrational motion of chain sections due to the weakening of cohesive forces in the polymer after its deformation. At the same time, the shift of the maximum of the bands skeletal vibrations turned out to be small (3 - 4 cm^{-1}), indicating an insignificant change in the force constant of these oscillations.

A similar conclusion was reached in [45], where the effect of chemical crosslinks on the packaging and properties of densely networked polydimethacrylates (PDMA) was studied. It was shown from the behavior of the bands of torsional skeletal vibrations that the transition from linear PMMA to cross-linked PDMA is accompanied by a significant broadening of these bands. Taking into account that the half-width of the bands of torsional skeletal vibrations is proportional to their amplitudes and anharmonicity coefficients, it was concluded that cross-linking led to an increase in the local free volume and a decrease in intermolecular interactions, i.e., changes that inevitably increase the local mobility of atomic groups in the network polymer. At the same time, the frequencies of skeletal vibrations of polymethacrylates chains in linear and cross-linked polymers differ insignificantly, which means that the loosening effect of crosslinks has little effect on the force constant of torsional vibrations, although it leads to a significant increase in their anharmonicity.

The discussed skeletal vibrations, undoubtedly, should largely determine the probability of relaxation transitions associated with large rotational displacements of chain sections, rather than limited torsional vibrations, which require, in particular, overcoming the barrier of internal rotation. Indeed, the FIR spectrum of pre-deformed PVC presented in [46] indicates a redistribution of intensities in the doublet band of torsional skeletal vibrations, which may be associated with a change in the conformational structure of the chain as a result of deformation.

The possibilities of studying conformational transformations and molecular mechanisms of relaxation transitions in polymers using long-wavelength IR spectroscopy are discussed in detail in the subsequent chapters of this review.

From other papers discussing skeletal vibrations in FIR spectra polymers, we point out the work of Chantry and coworkers [40], in which the spectra of ethylene-vinyl acetate copolymers (CEV) were studied. The most characteristic spectra of such a copolymer in comparison with the spectra of PVA and PE are shown in Figure 6.

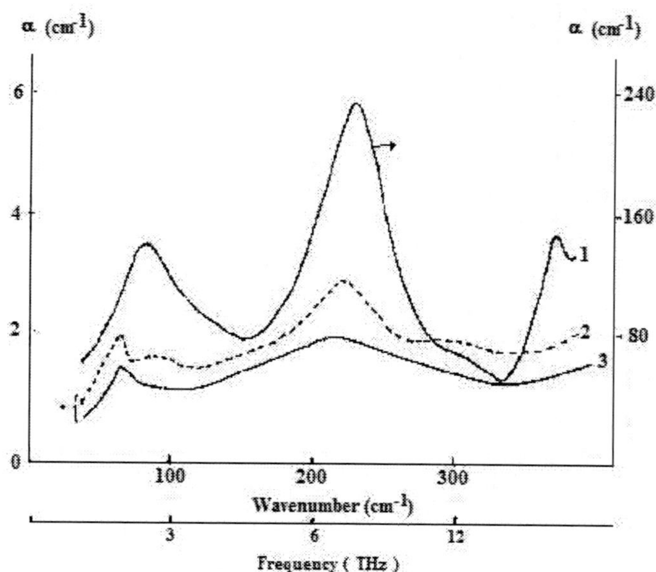

Figure 6. FIR spectra of PVA (1), copolymers of ethylene with vinyl acetate, 0.2% vinyl acetate (2) and PE (3).

Without touching the band of the liquid lattice with a maximum at 70–80 cm^{-} (see Chapter 4), let us pay attention to the absorption with a maximum at 232 cm^{-} in the PVA spectrum. For this polymer, there is no calculation of normal vibrations, but one can expect, judging by the conclusions of [47], that in such a polymer system, vibrations of the following three types will be active in the frequency range under consideration: (1) vibrational modes of the main chain, which become active in absorption in the presence of polar side groups, (2) modes of intramolecular vibrations the side groups themselves: with sufficient intermolecular interaction in macromolecules, (3) localized vibrations that capture a section of the chain in several repeating units, limited, for example, by such defects as conformational isomers. In this case, the absorption band due to such vibrations should be expected near the maximum of the density function of the vibrational state for the C-C chain. Judging by the intensity of the band at 232 cm^{-1} in the CEV with 0.2% vinyl acetate (Figure 6), which is much higher than could be expected, based on the low concentration of vinyl acetate in this copolymer, the latter of the three options is preferable. The retention of the frequency at the maximum of the same as in PVA indicates the participation of the acetate groups themselves in the absorption.

Chapter 3

Absorption by the Poly Mechanism in Low-Frequency IR Spectra of Polymers

3.1. Simplest Model for Libration-Driven Absorption

Theoretical predictions practically exclude the possibility of the participation of intramolecular modes in the formation of absorption bands in the region below 100 cm^{-1}. Thus, the lowest-frequency torsional skeletal vibrations, for example, around C – O or C – C bonds in peptide chains [49, 182], it can be made beginning only from 100-150cm^{-1}. At ν < 150 cm^{-1}, the absorption bands can be observed either due to lattice vibrations, just as the 73 cm^{-1} band in PE, or to intermolecular bonds vibrations - the H-bond for example.

It has now been established that the main contribution to the spectra of amorphous polymers in the region below 150 cm^{-1} is made by broadband absorption, which is sometimes background for narrow absorption bands of the crystal lattice. This absorption was first observed in the low-frequency IR spectra of PE, PP, and poly (4-methyl pentene-1) [2. 50] and was attributed to disorder-induced absorption in the amorphous regions of these partially crystalline polymers. However, a similar absorption band lying between the spectrum of relaxation losses and the spectrum of intramolecular vibrations is observed in all disordered media: liquids, solutions, glasses, liquid crystals, etc. In terms of position and integral intensity, this absorption corresponds to the spectrum of crystal lattice vibrations; therefore, it is often called a "liquid-lattice band" formed as a result of complete overlap of lattice bands upon amorphization of the structure [51-56].

Figure 7 shows the FIR spectra of (a) benzene, (b) cyclohexane, (c) acrylonitrile, and (d) chlorobenzene in their crystalline and liquid states. It can be seen that the spectrum of a molecular crystal in the region of 10–150 cm^{-1} with narrow bands of lattice vibrations degenerates into one broad absorption band on going to liquid. In terms of spectral parameters (intensity, half-width, and maximum position), this band is very similar to the absorption band in the FIR spectra of polymers with monomer units close in chemical structure, i.e., polystyrene (PS), polycyclohexyl methacrylate (PCGMA), polyacrylonitrile (PAN) and polychlorostyrene (PCS).

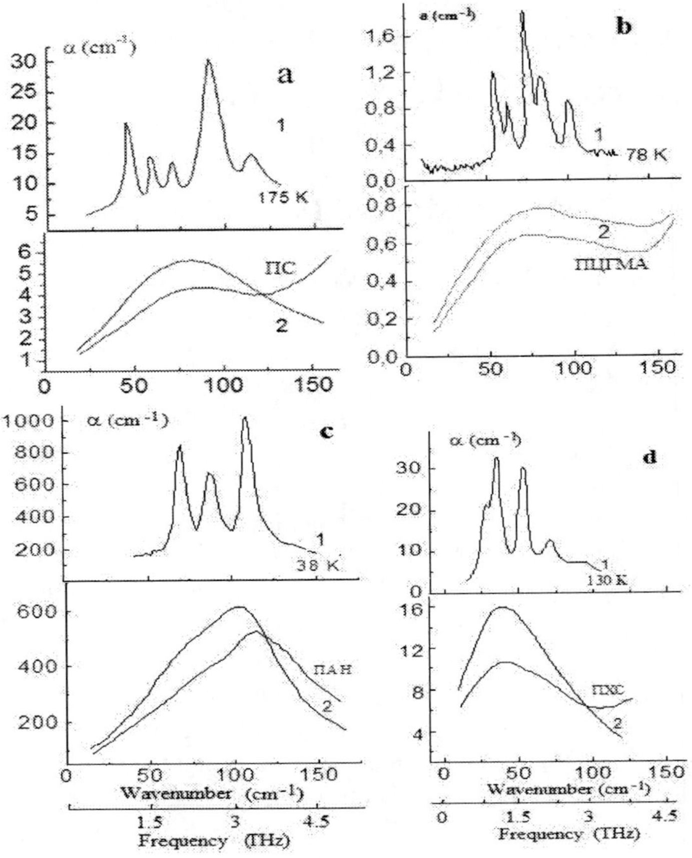

Figure 7. The "liquid lattice" band in the long-wavelength FIR spectra polymers PS, PCGMA, PAN, PCS and their monomers in crystalline (1) and liquid (2) states.

This correspondence indicates the general molecular nature of the discussed absorption in the spectra of low-molecular-weight liquids and polymers and allows it to be considered on the basis of the concepts developed for the interpretation low-frequency spectra of condensed media.

For the first time, the possibility of such absorption in the low-frequency IR range was proposed by J. Poley [100], who put forward the concept of librational movement of a polar molecule or dipole group in the field of intermolecular forces. Therefore, the "liquid lattice" band is also called "Poley librational absorption" or Poley-type absorption

The modern interpretation of spectra at low frequencies is based on the methods of molecular dynamics, using the statistical apparatus of the correlation functions of the dipole moment, defined as the Fourier transform of the spectral function. The technique is most successful as applied to microwave and FIR spectra of liquid systems, since these spectra reflect processes caused by short-lived orientational and translational fluctuations in a liquid.

One of the simplest models (and ways to obtain a specific analytical expression for the correlation function) currently used in the analysis of the "liquid lattice" band is the model of limited rotators or libration of a molecule in a potential field formed by its closest environment. The model is quite universal and makes it possible to describe both the absorption of Poley and the spectrum of dielectric losses in the terahertz range [52, 57, 74].

According to this model, the molecule takes part in torsional vibrations (libration) and rotation (reorientation). Its movement takes place within the potential well, which has the form $U(\varphi) = U_0 Sin^2 (\pi \varphi/2\xi)$, where U_0 is the depth of the potential well, ξ - is its half-width, φ - is the libration angle.

The model has only two independent "fitting" parameters that can be set from clear physical concepts: the librator lifetime between collisions, which depends on the height of the barrier limiting free rotation, and the amplitude of librational motion, determined by the configuration of the nearest neighbors. The motion occurs with a circular frequency $\omega_0 = \pi/\xi (Uo/2I)^{1/2}$, which at small angles of libration gives the frequency of the maximum band of the "liquid lattice" ν_{libr}= = $\omega_0/2\pi c$ = $(1/c\pi\varphi)(U_{libr}/2I)^{1/2}$, where I is the moment of inertia of the librator. The integral absorption intensity is determined by the expression: $A = \int_0^\infty \alpha(\nu)d\nu = \pi/3c^2 \sum \mu^2_z (1/I_x+1/I_y)$,

where $\alpha(\nu)$ is the absorption coefficient, and I_x and I_y are the moments of inertia of the molecule, the direction of which is perpendicular to the vector of the constant dipole moment μ_z , i.e., it is proportional to the square of the dipole moment and inversely proportional to the moment of inertia $I = (I_x + I_y)/2$. The linearity between the quantities $\alpha(\nu)$ and μ^2/I is shown in [53, 59, 102].

The role of the magnitude of the dipole moment of an oscillating molecule for the absorption intensity of Poley is discussed in [55] for nonpolar liquids, for which it is an order of magnitude smaller than in the case of polar ones. The intensity of this absorption in the spectra of nonpolar liquids is associated with induced dipole moments.

As an example of using the concept of the quasi-lattice structure of a liquid to interpret the FIR spectrum, we can cite the spectra of chlorobenzene at temperatures of 130 and 290 K [56]. Here, the broad absorption band ν_{max} = 45 cm^{-1} is attributed to the vibrational motion of the quasi-lattice of the liquid on the basis that in the spectrum of crystallized chlorobenzene at T = 130K, narrow absorption bands appear, corresponding to the lattice modes observed in the same frequency region (Figure 7d). According to [56], the experimentally observed broad band can be represented as consisting of a series of overlapping lines, the width of which is determined by the lifetime of the quasi-lattice state, which is close in order of the dielectric relaxation time. From the microwave spectrum of the liquid and from the measurements of the viscosity for chlorobenzene, it follows that the width of the overlapping lines corresponds to 10 cm^{-1}.

The concept under consideration is supported by the relative closeness of the integral absorption intensities for a liquid and a crystal in the region of lattice vibrations. Liquid-lattice absorption bands were found not only in liquids, but also in disordered crystals and glasses [54, 60].

3.2. Librational Absorption in Low-Frequency IR Spectra of Amorphous Polymers in the Range of 20 - 130 cm^{-1}

The manifestation of librational (by the Poley) absorption in the FIR spectra of partially crystalline and amorphous polymers [2, 31, 61] represents information that is fundamentally different from that obtained from the mid-IR spectra, since it is characterized by a greater role of intermolecular dynamics.

It was noted that the conditions for observing this absorption in the spectra of solid polymers are more favorable than in the spectra of liquids. In liquids, relaxation times are of the order of 10^{-11} sec and the manifestation of Debye relaxation occurs in the region of 1-10 cm^{-1}, that is, there is an overlap with the band of the liquid lattice. In polymers, the relaxation times are generally higher (usually 10^{-9} sec) and the loss peaks, caused by Debye relaxation, are shifted to the radio frequency region, increasing the probability of separating these losses from absorption by the Poley mechanism.

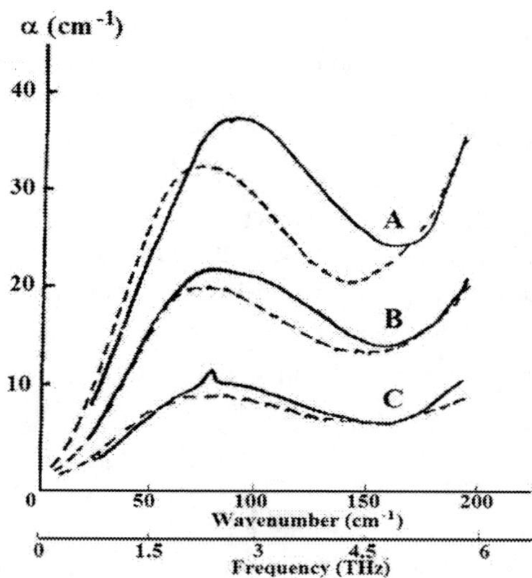

Figure 8. FIR spectra of a copolymer of ethylene with vinyl acetate at T = 295K (dashed lines) and T = 78K (solid lines). Curves: (A) -10%, (B) -6%, (C) -3% vinyl acetate.

Copolymers of ethylene with 3, 6 and 10% vinyl acetate [40, 61] have proved to be successful objects for studying liquid lattice absorption in polymers. The intensity of the lattice mode characteristic of the PE spectrum at 73 cm^{-1} in the FIR spectrum of these copolymers decreases, but additional background absorption appears, attributed to absorption by the Poley mechanism (Figure 8). It is interesting that the direct proportionality between the absorption coefficient and the concentration of polar molecules appears even when the latter are chemically bound with a non-polar environment, which also coincides with that observed in polar liquids. The assignment is confirmed by the shift of the frequency of the maximum to the low frequencies with decreasing temperature. Since the band does not decrease in intensity in this case, it cannot be attributed to the difference mode, and the large half-width of the band does not allow associating it with any intramolecular vibration mode in the acetate group.

Another example of poly absorption is a broad absorption band in the PTFE spectrum centered at 20° C at 50 cm^{-1}. In the FIR spectrum of this highly crystalline polymer at low temperatures (Figure 3c), in the region of this peak, several narrow bands are visible, which are close in frequency to those predicted for crystal lattice vibrations: translational at 46, 55 and 70 cm^{-1} and rotational at 55 cm^{-1} [62]. The authors in [63] interpreted these bands from the standpoint of the manifestation of the conformational disordering of chains. However, thermal treatment of PTFE, which significantly reduces its crystallinity, did not lead to an increase in the intensity of these narrow bands due to an increase in conformational disordering, but to their decrease and the appearance of simultaneously the above-mentioned background absorption, i.e., to the absorption of Poley.

In the two available small reviews on the FIR spectra of polymers (1, 2), only a passing mention is made of the possible presence of Poley absorption in the FIR spectra of amorphous polymers PS and PMMA at $\nu < 150$ cm^{-1}. Some more information can be obtained from Raman spectra. For example, in [36], the band at ~ 60 cm^{-1} in PS was attributed to the librational motion of benzene rings. A similar assignment was made for the absorption band with a peak in the 40-65 cm^{-1} region for substituted PSs, showing a decrease in intensity with increasing mass side group (see Applications 9.1).

For the absorption band at 65 cm^{-1} in the Raman spectrum of the PMS, a doublet contour was observed, probably due to the presence of two conformations in the basic chain.

Similarly, in the Raman spectrum of amorphous PMMA, the libration band lies at 70 cm^{-1} [64], and in PS, at 80 cm^{-1} [65]. In addition, in the Raman spectra of PS, PMMA, PC, and substituted PS [36, 64-66], absorption is also observed in the region $\nu \leq 50$ cm^{-1}, which the authors attribute to the manifestation of the translational motion of the chain.

Let us turn again to the FIR spectra of PMMA, PMA, PVA, and PVC in the range $\nu \leq 130$ cm^{-1}. FIR spectra of isotactic, syndiotactic and atactic PMMA, showing broad doublet bands at 97 and 123 cm^{-1}, 80 and 100 cm^{-1}, 75 and 95 cm^{-1}, respectively, are presented in [38]. The bands were attributed to the torsional vibrations of the side groups, and their doublet character is associated with the influence of the rotational isomerism of the ether group. Similar bands at 83 cm^{-1} in the PMA spectrum and at 85 cm^{-1} in the PVA spectrum have no structure in agreement with the absence of rotational isomerism in these polymers. It is noted that highly symmetric torsional vibrations of methyl groups should not appear in the spectra at these frequencies.

Low-frequency modes in the PVC spectrum were first observed by Krimm et al. [67] at 67, 89, and 180 cm^{-} and were correlated with the calculation based on the normal coordination analysis [68, 69], which predicted absorption bands at 54 and 117 cm^{-1} due to mainly by torsional vibrations and a band of 127 cm^{-1}, corresponding to wagging mode of the C-C1 group. Somewhat later, these authors [70], using a more realistic model that takes into account intermolecular interactions of the C – H ... Cl – C type, predicted the absorption at 29, 64, and 90 cm^{-1}. The band at 29 cm^{-1} was attributed to the deformation mode of the C-Cl ... H group, at 64 cm^{-1} to the torsional-deformation vibration, and at 90 cm^{-1} to the torsional vibration of the main chain. Later, in [71], temperature measurements of the PVC spectrum were carried out, confirming the assignments, but the band at 64 cm^{-1} was assigned to the lattice mode, and the band at 90 cm^{-1} was attributed to absorption by the Poley mechanism, which is confirmed by the nature of the dependence of the spectral parameters of this band on the degree of ordering and syndiotacticity of PVC.

Systematic study of the liquid lattice band in polymers was carried out in [30, 31, 72, 73]. The objects were polymers of different structures and predominantly amorphous (PS, PMMA, PVC, PCS, PAN, PC, etc.), and librational absorption was considered as a fairly general phenomenon. A feature of these works was that to assign the considered absorption, the authors used the results of the analysis of the FIR spectra of low-molecular-weight liquids, which closely correspond in chemical structure to the monomer unit of the polymer.

Comparison of the FIR spectra of amorphous polymers with the literature data on the spectra of liquids made it possible to establish that the absorption at frequencies below 150 cm^{-1} is of the same nature. The main contribution here comes from the libration of the molecule (in a liquid) and a monomer or polar side group in macromolecules.

This connection is well traced, for example, in the case of PCS, which has a chemical structure of a monomer unit similar to chlorobenzene. Figure 7d shows a broad band at 47 cm^{-1} in the PCS spectrum, corresponding to the band in liquid chlorobenzene at the same frequency that is attributed to the librational motion of chlorobenzene [56]. The spectrum of crystalline chlorobenzene contains sharp bands at 37, 55 and 70 cm^{-1}.

Absorption with a maximum at 80 cm^{-} in the PS spectrum (Figure 7a) corresponds to the same absorption in the spectrum of liquid benzene at 75 cm^{-1}, which is attributed to libration [76].

The spectrum of lattice vibrations of crystalline benzene in this region contains bands at 65, 74, 92, and 116 cm^{-1} of close total intensity.

The FIR spectra of crosslinked PS [77] and polypentafluorostyrene PPFS [78] are indicative. In the former, at a high concentration of the crosslinking component of divinylbenzene (DVB) in the styrene-DVB copolymer, the 80 cm^{-1} band shifts by 10 cm^{-1} to high frequencies due to the greater inhibition of the libration movement of benzene rings, which are chemically bonded immediately with two chains. In the case of PPFS, the liquid lattice band is shifted, on the contrary, to low frequencies (Figure 9) in accordance with the dependence $v \sim (E_{coh})^{1/2}$ due to the greater moment of inertia of the fluorine-substituted benzene ring. Figure 9 (curve 3) shows a broad band at 23 ± 4 cm^{-1} in PPFS, corresponding to the absorption band of pentafluorobenzene, the structure of which is similar to the monomer unit of PPFS.

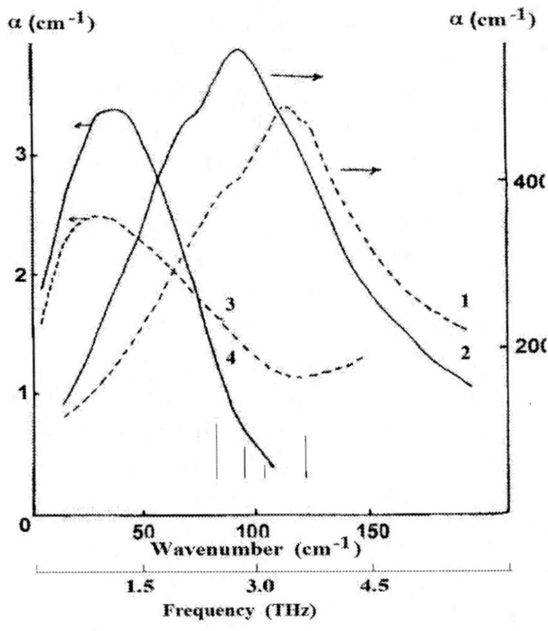

Figure 9. The band of the liquid lattice in the FIR spectra of PAN (1); liquid acetonitrile (2); PPFS (3) and pentafluorobenzene (4). The vertical lines show the position of the stripes in spectrum of crystalline acetonitrile.

As follows from (71), an absorption band with a maximum at 90 cm^{-1} in the FIR spectrum of PVC due to the libration of groups containing a chlorine atom. In this case, the authors consider the absorption near 64 cm^{-1} as a lattice mode.

A similar assignment can be made for PVF with a structure similar to PVC and having an absorption in this region with a maximum of 88 cm^{-1} and a shoulder of 66 cm^{-1}.

In the FIR spectrum of PAN (Figure 4 and Figure 9) in the region of 25–150 cm^{-1}, the absorption by the Poley mechanism should include a band with a maximum at 127 cm^{-1} and a shoulder at 85 cm^{-1}. According to the authors of (41), the band at 127 cm^{-1} corresponds to torsional vibrations CN group of the PAN. However, in the spectrum of liquid acetonitrile (with the chemical structure close to the structure of the PAN monomer unit, the broad band has a maximum at 100 cm^{-1}, which in [79] was attributed to the librational motion of acetonitrile dimers. In the same spectral region, absorption in the spectra of simple nitriles [80] and acetonitrile at low temperatures [81] was associated with the libration of dipole – dipole complexes of various configurations. Thus, the discussed band in the PAN spectrum characterizes the torsional vibrations of the monomer units of the polymer and the intermolecular vibrations of their complexes.

As noted above, in the FIR spectrum of PMMA (Figure 21), to torsional vibrations of the side group or, at least, a unit of motion not exceeding monomer unit, include the absorption band with a maximum at 90 cm^{-1} [38]. Its complex contour can be associated with the existence of two rotational isomeric states of the ether group. A certain contribution to the absorption, according to the calculation [21], is apparently also made by out-of-plane skeletal vibrations: $\nu_5 = 73$ cm^{-1}.

An interesting fact was that the FIR spectra in this region did not undergo fundamental changes with the transition from PMMA to comb-like polymers of this series, such as PBMA, POMA, PDMA, and PCHMA, that is, with an increase in the polymethylene side group (Figure 10) [39]. Although due to the big moment inertia of such groups, one might expect that the maximum of librational absorption would shift significantly to lower frequencies. This suggests that non-polar polymethylene side groups attached to chains through movable “oxygen hinges” behave quasi-independently and do not make a significant contribution to this absorption. Libration movement in this series of polymers is carried out mainly within -CH_3 – C - C = O group.

This is confirmed by the closeness of the intensities of the discussed absorption in the spectra of PBMA, POMA, and PDMA, if we conditionally recalculate the absorption coefficients only for the number of absorbing centers. The presence of non-polar side polymethylene chains affects only a slight decrease in the amplitude of the librational movement, which shifts the maximum of the libration band to higher frequencies (from 95 cm^{-1} to 105 and 110 cm^{-1}) and a decrease in intensity relative to the band in the PMMA spectrum.

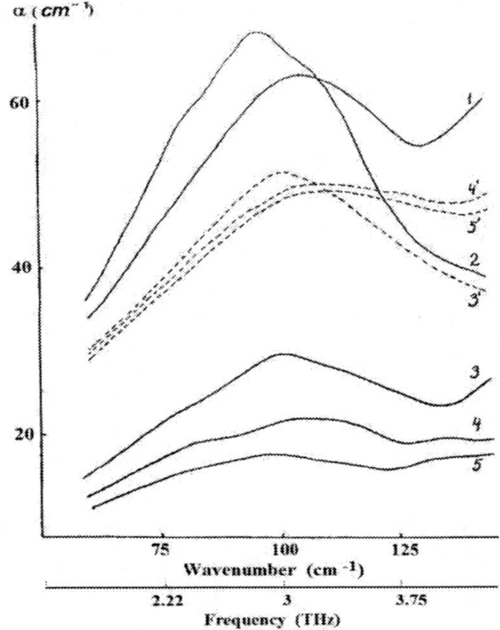

Figure 10. The band of the liquid lattice in the FIR spectra of PCHMA (1), PMMA (2), PBMA (3), POMA (4), and PDMA (5). In 3, 4, and 5, the band intensities were calculated without taking into account the contribution of polymethylene side groups.

Finally, the libration band changes its parameters relatively little with a change in the degree of polymerization. In experiments with MMA oligomers [72], absorption centered at 95 cm^{-1} was also observed for oligomers with n = 2, 7, 9, and 50 units. Moreover, the spectrum of the oligomer with n = 9 already practically coincided with the spectrum of the polymer (Figure 11).

The same picture was observed for PMS oligomers (82) (Figure 12a). The torsional vibration band of the phenyl group in the spectrum of this polymer has a doublet character with maxima at 75 and 95 cm^{-1}. This is especially clearly seen in the Raman spectrum of PMS [83] and is explained by the influence of the conformation of the main chain on the movement of the phenyl group. It was shown in [83] that in the *tttt* conformation, the vibration plane of the benzene ring corresponds to the bisector of the bond angle of the chain, while in the *ttg* + *g* + conformation; it vibrates in the bond angle plane. These two stable conformations give, obviously, two distributions of torsion frequencies in the PMS.

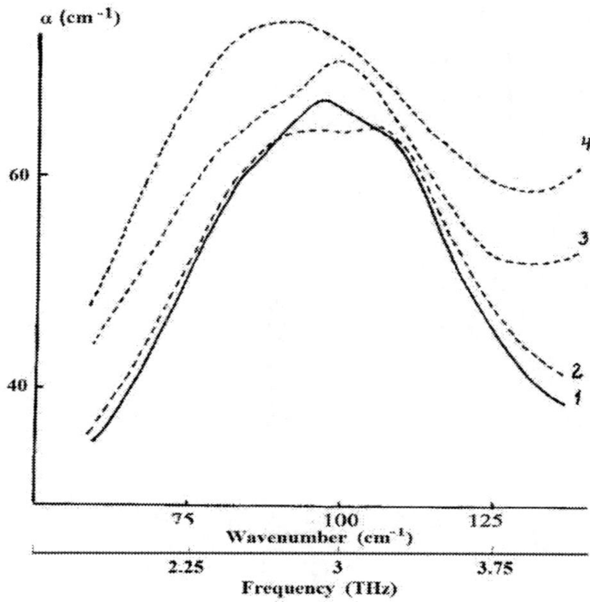

Figure 11. Liquid lattice band in the FIR spectra of PMMA (1) and oligomethacrylates with $n = 50$ (2), $n = 7$ (3) and $n = 2$ (4).

Localized librational movement of atomic groups in oligomers and polymers, realized in the absence of significant side groups can be illustrated with data obtained on the PC (82) and PDMS [84]. In PC and oligocarbonates, the librational movement within a sector of 20 degrees determines the absorption centered at 75 cm^{-1} (Figure 12b). The same assignment was made for the 80 cm^{-1} band in the Raman spectrum of PS [163]. With an increase in the degree of polymerization, the 75 cm^{-1} band in the PS spectrum practically does not change in frequency: only its peak intensity, which is increased in the case of a dimer and a trimer, decreases noticeably. At the same time, in the PDMS spectrum with n = 2, 50, and 1000 there is a similar absorption band, due to torsional vibrations of monomer units, as n increases. Apparently, this can be explained by the increased flexibility of PDMS macromolecules and the greater independence of the libration motion of dipoles in this polymer. The same can be concluded from the position of the maximum of the libration band. Its shift to lower frequencies with increasing molecular weight, as well as the case of PC is insignificant, showing that the parameters of the librational movement of the molecule in the liquid and the monomer in the macromolecule do not differ much from each other. And this will be an additional indicator of the local nature of the observed mode, although the large bandwidth in the spectra of PS and PDMS still indicates a significant disturbance of this motion by the interaction with the environment in the case of a polymer.

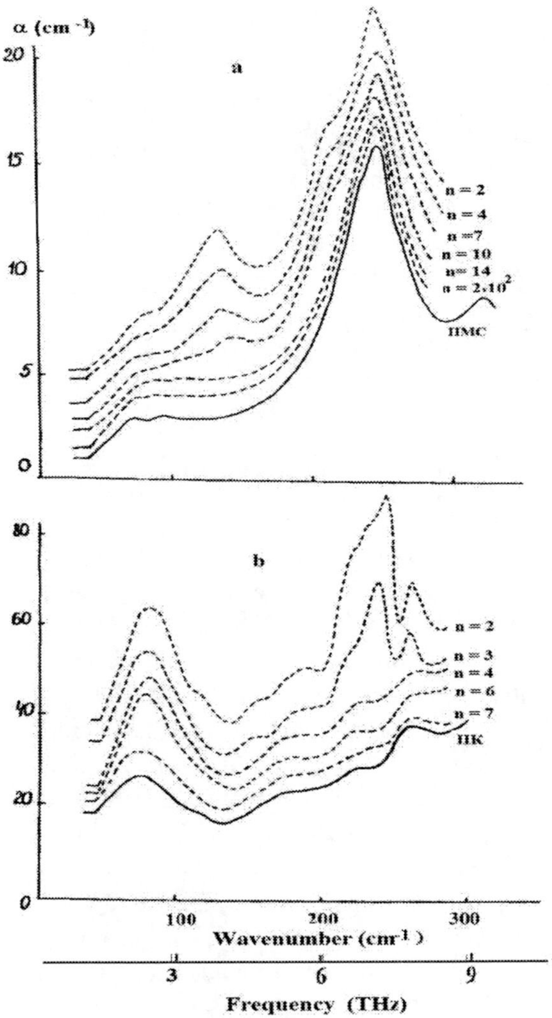

Figure 12. (a) FIR spectra of PMS and its oligomers with $n = 4, 7, 10, 14$ and $2 \cdot 10^2$; (b) FIR spectra of PC and oligomers with $n = 7, 6, 4, 3$ and 2. The spectra of oligomers are shifted along the ordinate.

3.3. Relationship between the Parameters of Librational Absorption and Molecular Characteristics of Polymers

In [30, 31], the spectral parameters of low-frequency absorption bands, attributed to the absorption by the Poley mechanism due to the librational motion of polar groups, were compared with the molecular characteristics of the studied polymers.

At the same time, they proceeded from the following provisions:

1. The frequency of torsional vibrations (libration) ν_{libr} is proportional to the value $(U_{libr}/2I)^{1/2}$, where U_{libr} is the height of the potential barrier for libration, and I is the moment of inertia [57]. This relationship is indeed observed for a large number of simple liquids [53] and molecular crystals. In the latter, the frequency of rotational vibrations, according to [85], is related to the lattice energy and structure through the proportionality $\nu_R = (2\Delta H_s/I)^{1/2}$, where ΔH_s correspond to the lattice energy component, and I is the main moment of inertia. That is, the librational frequency is determined by intermolecular forces, which can be measured directly in a molecular crystal as the heat of stimulation.
2. By analogy with liquids [53], a correlation is assumed between the coefficient absorption for the libration band $\alpha\ (\nu)$ and μ^2_{eff}/I, where μ_{eff} is the effective dipole moment of the monomer unit (librator).
3. There should be a relationship between the value of the barrier for libration U_{libr} and the cohesion energy E_{coh}, which characterizes the molecular interactions in the polymer.

The μ_{eff} values in the calculations for PVC, PCS, PMMA, PBMA and PVA were taken from reference manuals [86, 87]. According to [87], the interaction of dipoles in neighboring units of the macromolecule leads to the fact that μ_{eff} of polymers is less than the dipole moments of the corresponding monomeric liquids μ_0: the correlation parameter is $g = \mu^2_{eff}/\mu^2_0 = 0.7 \pm 0.1$. This was taken into account in calculating μ_{eff} for PVF, PAN, and PS. When calculating the moment of inertia $I = 2MR^2_{eq}/5$, where M is the molecular weight of the librating unit, and R_{eq} is its equivalent radius obtained from the van der Waals volume [88], the latter was assumed to be spherical.

As can be seen from Figure 13a, for carbon chain polymers as well as for liquids [53] it is observed, confirming such an assignment, the proportionality between α *(ν)* and the value of μ^2_{eff}/I. The relationship between ν_{libr} and E_{coh} (Figure 13b), reflecting the relationship between the potential barrier to libration and the level of molecular interactions in the polymer, is also presumably established.

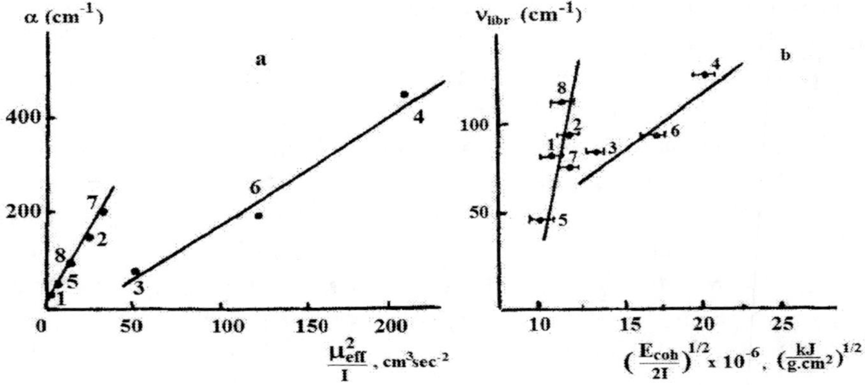

Figure 13. (a) Intensities and (b) frequencies of the maxima of the bands of the liquid lattice in the FIR spectra 1, PS; 2, PMMA; 3, PVC; 4, PAN; 5, PCS; 6, PVF; 7, PVA; 8, PBMA.

For PVC, PVF, and PAN, the experimental points have different relationships between α (ν) and the value of μ^2_{eff}/I and between ν_{libr} and E_{coh}. This deviation can be due to the large μ_{eff} of these polymers, as well as due to their specificity. As you know, macromolecules PVC, PVF and PAN have increased rigidity, namely, their Kuhn segment contains 10-12 monomer units, while in other polymers considered here it does not exceed 6-8 units. Probably, this leads to the unification of the librational movement of neighboring units, and thus the parameters of the librational movement for such polymers should differ from the parameters obtained for individual monomer units.

The revealed correlations indicate that the FIR spectra of amorphous polymers in the region below 150 cm^{-1}, like the spectra of liquids in the same range, exhibit absorption due to limited torsional rocking (libration) of monomer units; the characteristics of this absorption are determined by such molecular parameters as the effective dipole moment of the monomer unit, its geometry, and the cohesion energy. Consequently, this section of the FIR spectrum carries information about the dynamics of monomeric units, depending on their structure and non-chemical interactions with the environment.

Thus, the “liquid lattice” band (absorption by the Poley mechanism) in the FIR spectra, in addition to the characteristic inherent in the spectrum of intramolecular vibrations in the mid-IR range, carries nontrivial information about intermolecular mobility and interactions in condensed media. The ability to directly analyze the influence of molecular structure and cohesive forces on the local (and segmental) dynamics of macromolecules opens up new approaches to the development of topical issues in polymer physics.

Chapter 4

Low-Frequency IR Spectra and Molecular Interpretation of Relaxation Processes in Amorphous Polymers

4.1. Submillimeter and Millimeter Spectra and Dielectric Relaxation

So, we see that when analyzing FIR spectra, as well as mid-IR spectra, certain frequencies are interpreted at which resonant absorption of radiation by an oscillating system of atoms or molecules occurs. However, in polymers, at certain frequencies and temperatures, processes appear that reflect configurational and other permutations, which also lead to the dissipation of the energy of the applied electromagnetic field.

Various techniques of dielectric or mechanical spectroscopy make it possible to record this in the form of relaxation absorption: the α-loss peak directly in the vicinity of T_g (glass transition temperature) and β-, γ- and δ-peaks at temperatures below T_g [89, 90].

It is important to emphasize the physical difference between relaxation and resonance movement, since there are often misunderstandings due to the use of the same terminology to describe physically different phenomena. Let us consider a short segment of a polymer molecule, which, for steric reasons, has a definite configuration and, therefore, a quite definite direction of the dipole moment. The segment can absorb IR radiation, exciting any of its normal optically active modes. As a result, we have resonant absorption bands with maximum frequencies corresponding to vibrational frequencies. The frequencies at which the absorption is greatest are, to a first-order approximation, independent of temperature since they are determined only by the shape of the potential well surface, but not by the distribution of molecules over possible energy levels. Now let us consider an irreversible jump of a segment from an initial position to another: the orientation of the dipole moment will now be different, and if the jump is accompanied by the resistance of the medium, energy will be absorbed from the electromagnetic field and, as a result, a relaxation type of absorption. Barriers opposing segment reorientation will be temperature sensitive and thus relaxation modes will also change with temperature.

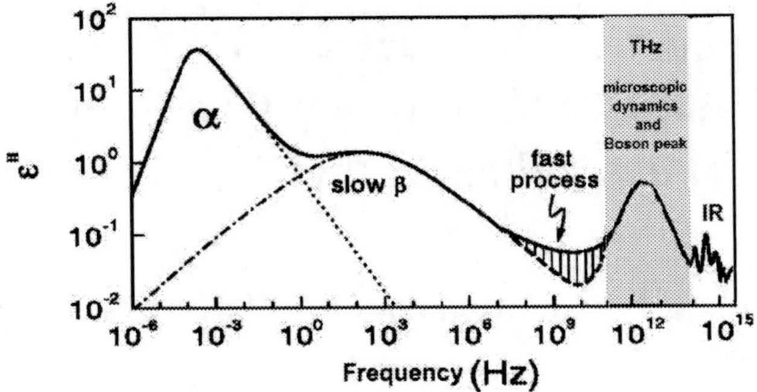

Figure 14. Schematic representation of various relaxation and resonance processes observed in disordered condensed media by broadband dielectric and optical spectroscopy. α: primary relaxation (glass transition), slow β: secondary relaxation (Johari-Goldstein process [91, 102]), fast process: a process similar to fast β [103], microscopic dynamics and Boson peak [104, 105]: peak of the density of vibrational states, including libration, *IR*: intramolecular infrared modes. Shading marks the frequency window that is available to FIR, Raman, and terahertz spectrometers.

When measuring relaxation spectra, it is desirable to use the widest possible frequency range. Modern methods of dielectric spectroscopy allow working in the range from 10^{-6} to 10^{15} Hz (Figure 14) [101]. Figure 14 shows that in the submillimeter range, covered by optics and radio engineering, relaxation and resonance processes coexist in such a way that their distinction becomes nontrivial. Libration-mediated Poley absorption and relaxation spectrum are two inseparable phenomena, connected by a single process of orientational dynamics of molecules in a condensed state, and therefore they should be considered not independently, but from a general point of view. This method of studying the dielectric spectra of condensed media in a wide frequency band is sometimes also called 0 - THz spectroscopy [74].

0 - THz spectroscopy is currently one of the most informative methods for studying the dynamics of molecules in liquids [60]. This is due to the fact that the characteristic frequencies of spectroscopically active molecular movements in a liquid correspond precisely to the terahertz (low-frequency IR and Raman) ranges and appear in such details that are difficult to obtain by other spectral methods (mid-IR spectroscopy. NMR, Rayleigh scattering, etc.).

These possibilities of 0 - THz spectroscopy in the case of polymers are well illustrated in Figure 15, which shows the Raman spectrum of PMMA in the range 0.001 - 10 THz [108], reduced in coordinates $I(\nu)\,[n(\nu)+1]^{-1}$, where $I(\nu)$ is the intensity spectrum of scattering, and $n(\nu) = = [exp(\hbar\nu/k_BT) - 1]^{-1}$ is the Bose factor.

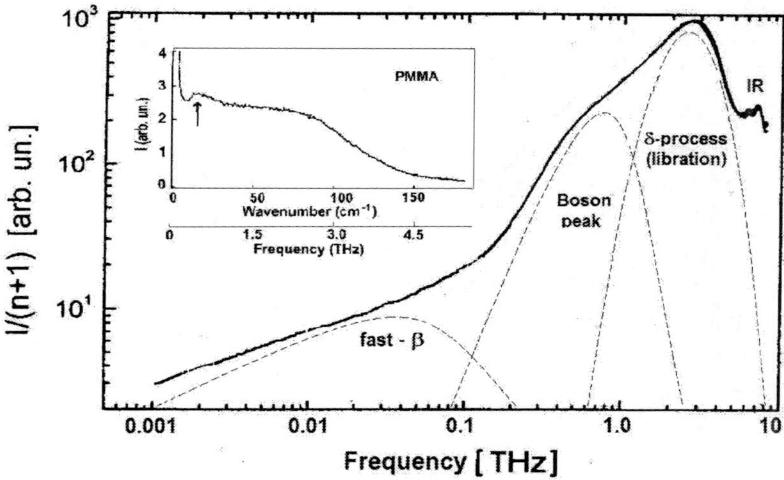

Figure 15. Reduced Raman spectrum of PMMA. The inset shows the experimental spectrum of PMMA at room temperature. The arrow shows the position of the boson peak in the spectrum.

Representation of Raman - spectra in coordinates $I(\nu)\,[n(\nu)+1]^{-1}$ has the advantage that it allows direct comparison with spectra of dielectric losses in coordinates $\varepsilon''(\nu)$. In addition, since $\varepsilon''(\nu) \sim n(\nu)\,k(\nu)$ [1], where $n(\nu)$ is the refractive index, and $k(\nu)$ is the damping index (extinction coefficient) related to $\alpha(\nu)$ - the absorption coefficient in long-wave and terahertz IR spectra by the ratio: $k(\nu) = \alpha(\nu)/4\pi\nu$, low-frequency IR spectra in coordinates $\alpha(\nu)/\nu$ can be compared not only with spectra of dielectric losses, but also with Raman spectra in coordinates $I(\nu)\,[n(\nu)+1]^{-1}$.

Figure 15 shows that the low-frequency Raman spectra of polymers (in this case, PMMA) carry information not only about resonance processes (intramolecular vibrations and libration), but also about collective vibrational excitations represented in the experimental spectra by the "boson peak" (BP) and the inflection point at the frequency of the BP maximum in the

[1] More detailed information in Application 9.2.

(I (ν)/[n (ν) +1])$^{-1}$ spectra. And also, about the relaxation dynamics, represented here by the reorientation of the C – O – C plane of the side group of this polymer (fast *β*-relaxation in the terminology of [106]). More detailed information on BP and fast *β*-relaxation in polymers is presented below. Another example of studying the dielectric spectrum over a wide frequency range (including IR) also shows that n-alkanes (C5 - C14) have a wide dispersion in the microwave region, accompanied by dispersion in the FIR [94]. Frequency the tan δ dependence (Figure 16) here has two clearly separate peaks:

1. High-frequency absorption corresponding to resonant Poley-type absorption (γ - process following the notation [60] -the counterpart of δ-relaxation in polymers) and it is due to short time and small-angle torsional oscillations of a dipoles within a one potential well. The contour and position of this absorption peak are independent of the number of carbon atoms in the chain.

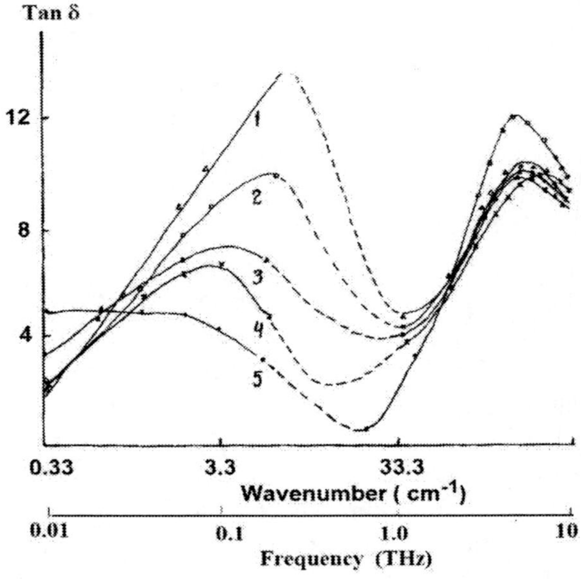

Figure 16. Dielectric loss *(tan δ* × 104) vs *ν* (THz) or (cm^{-1}) for *n*-alkanes: pentene (1), hexane (2), heptane (3), nonane (4), and tetradecane (5).

2. And broadband low-frequency absorption (in the region of 10^9 - 5 · 10^{11} Hz), corresponding to Debye relaxation (β - process) as a result

of slow diffusion of dipoles through large angles from one potential well to another. The maximum of this absorption shifts to lower frequencies with increasing chain length. Based on its position, the Debye relaxation time $\tau_d = 1/2\pi c\nu_{max}$ was obtained; it grows with the number of carbon atoms in the chain.

In PE, the peak of Debye relaxation (reorientation of segments of the main chain) appears at a frequency of 10^{11} Hz [61]. On the temperature dependence of mechanical losses (at kHz), it corresponds to a peak at 170 K [95]. It is clearly visible in Figure 2f, which shows the full spectrum of dielectric losses of the PE, including the FIR range with resonances at 73 cm^{-1} (mode $\nu^a{}_5$ $(\theta = \pi)$) and absorption centered at 200 cm^{-1}, due to skeletal (torsional and bending) vibrations of the main chain.

The absorption of Poley in the region of 100 cm^{-1} is due to the libration movement of monomeric units of the chain and characterizes the δ-process in this non-polar polymer (see below). Dielectric losses in PE in the microwave region are due to the presence of polar impurities, end groups, folds, and branch points.

Completed by Buckingham et al. [96] measurements on a copolymer of ethylene with PP show an increase in losses in the region 10^8–10^9 Hz compared to PE due to the higher polarity of PP. Unlike the spectrum of losses of PE, the spectrum of the copolymer contains also *α* peak at 10^4 Hz at room temperature. The effect of oxidation is described in [95], where it was shown that the peak at 10^8 Hz in the spectrum of dielectric losses of PE grows with its oxidation.

The works [2, 78, and 107] compare the spectra of dielectric losses in PE, PP and poly (4-methyl pentene-1) in the region of the low-frequency wing δ - relaxation. These data, presented in a spectroscopic manner as the spectrum of the absorption coefficient $\alpha(\nu)$, in cm^{-1}, are shown in Figure 17. As can be seen, in this representation, the manifestation of relaxation losses is expressed not by a curve with a maximum, as in the spectrum $\varepsilon''(\nu)$, but by a monotonic increase in $\alpha(\nu)$, with an increase in frequency with a noticeable increase in the slope for more substituted chains. Since $\tan\delta$ at the peak of the β-process can increase by an order of magnitude, for example, during oxidation, one can expect its contribution to the intensity of the low-frequency wing of the FIR spectra of these non-polar polymers.

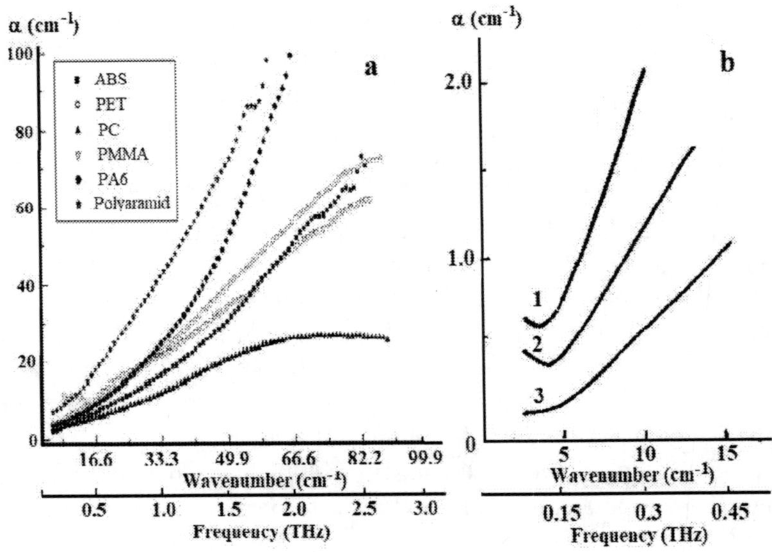

Figure 17. a, b. FIR and terahertz absorption spectra. a: ABS, PET, PC, PMMA, PA6 and Polyaramid and b: vinyl acetate-ethylene copolymers (CEV) with (1) 10%, (2) 6% and (3) 3%.

In a more polar copolymer of ethylene with vinyl acetate (CEV), it is possible to separate observation of the manifestation of relaxation absorption and absorption by the Poley mechanism. As discussed earlier, the broad absorption band observed for this polymer (Figures 6 and 8) is a liquid lattice band, i.e., absorption by the Poley mechanism. In the spectrum of dielectric losses at these frequencies, obviously, there should be a loss peak corresponding to δ - relaxation, as is observed in simple molecular liquids [60, 93], while β - relaxation is expected for this copolymer at 10^9 Hz. Almost all spectra in the submillimeter range are extrapolated to the origin, but this is not the case in the spectra in Figure 16a. Below 5 cm^{-1}, the spectrum shows a change in slope, indicating another absorption band lower in frequency - pseudo-relaxation. Thus, if the absorption below 5 cm^{-1} is a high-frequency wing of the β-process, there is a real separation of it from absorption by the Poley mechanism.

In the spectrum of dielectric losses of a low-polarity PS (Figure 18) below 10^{10} Hz at room temperature [18] on this scale there are no clearly expressed relaxation peaks of losses due to α or β processes. The only peak at $\nu \approx 2 \times 10^{12}$ Hz corresponds to absorption at 80 cm^{-1} caused by the libration of benzene rings, i.e., δ - relaxation. In the spectrum of mechanical losses, it corresponds to a peak at 46 K (1 kHz) with an activation energy of 12 kJ/mol [97].

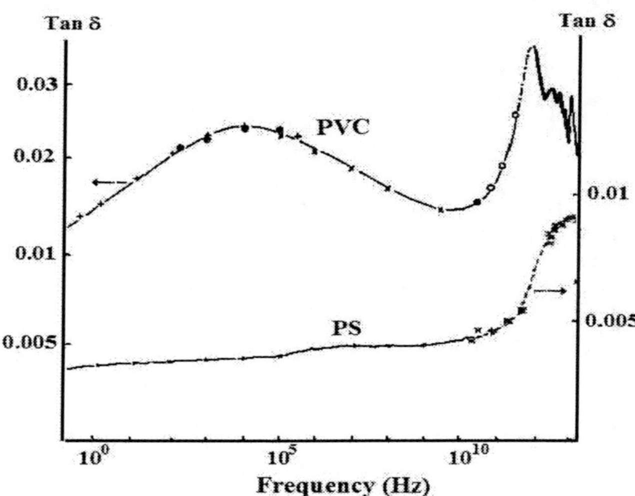

Figure 18. Spectra tan δ for PS and PVC at room temperature.

Figure 18 also shows the spectrum of dielectric losses of PVC at room temperature in a wide frequency range together with the data on the FIR absorption in this polymer in the range of 3.10^{10}-$0.5.10^{13}$ Hz, taken from [18]. The peak loss at 10^4 Hz is due to *β*-relaxation, which characterizes the local segmental motion in the main circuit. In the spectrum of mechanical losses, this peak is located at T = 230-250K, (1 Hz) [99]. The most high-frequency losses in the dielectric spectrum correspond to the low-frequency wing of the absorption band with a maximum at 90 cm^{-1}, which corresponds to librational oscillations in the main circuit, i.e., correspond to the *δ* - process in this polymer.

With an increase in the polarity of the librating atomic group in substituted PSs, dielectric losses increase (Figure 19 and Application 9.1). In (78), observations were performed on polypentafluorostyrene (PPFS), in which aromatic nuclei contained 5 fluorine atoms. As expected, with an increase in the side group mass, the frequency of the maximum shifts to 3 cm^{-1} (~ 10^{11} Hz), while the peak loss in comparison with the PS increases.

In polytrifluorostyrene (PTFS), where the hydrogen atoms in the main chain are replaced and the benzene ring is not touched, the maximum loss occurs at 40 cm^{-1} (compare with 70 cm^{-1} in PS), but the losses are several times greater than in PS. This indicates that the δ-process in polymers reflects the libration of the side groups, but at the same time properties of the backbone are also important.

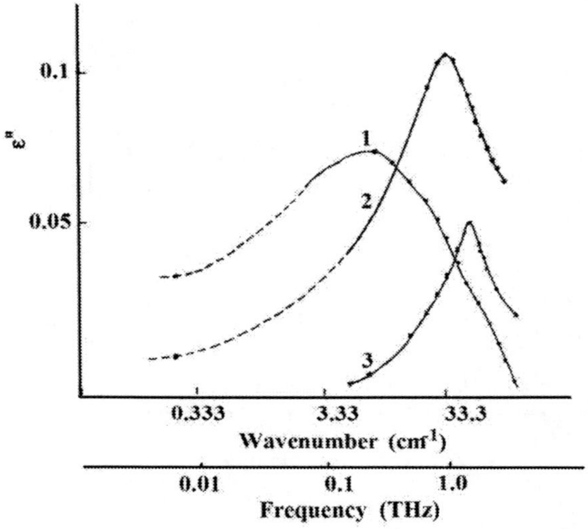

Figure 19. Dielectric loss spectrum ε '' vs ν (cm^{-1}) or υ (THz) for PPFS (1), PTFS (2) and PS (3).

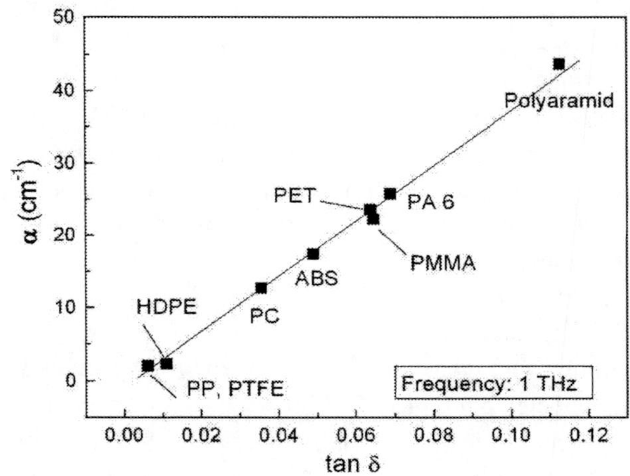

Figure 20. Relationship between absorption coefficient α (cm^{-1}) and loss tangent (*tan* δ).

Among recent works on the study of dielectric loss spectra in polymers, one should dwell on works performed using terahertz spectroscopy (THz-TDS) [107, 109 and Application 9.3], which allows one to simultaneously obtain the spectra of loss and refractive index. From the obtained spectrum of the refractive index, the absorption coefficient and data on the dielectric constant of the material can then be obtained. Figure 20 shows the relationship between the absorption coefficient and the loss tangent for the eight polymers studied in [107]). This result indicates that the main mechanisms of absorption of terahertz radiation in polymers, which determine dielectric potentials, are collisions and vibrations of macromolecules.

In (110), the terahertz spectra of 16 polymers, such as PMMA, PS, PC, PET, and polyacrylates of the phenoxy ethylmethacrylate (PEM) type were studied. All experimental data show a characteristic broad absorption peak at a frequency near 2.5 THz, accompanied by a noticeable negative dispersion (on average, $dn/df = -0.01$ THz^{-1}) at lower frequencies. The authors believe that the mechanism of this absorption is associated with relaxation and an amorphous state.

Using the modified Debye model to describe the absorption and refractive spectra of the studied polymers in the 0.1–3 THz region, they attempt to elucidate the influence of flexibility, polymerization, degree of crystallinity, and the number of monomeric end groups on the dielectric properties of the studied polymers. Note that in this review (Section 5.2), this anomalously wide absorption peak in the terahertz IR spectra of polymers is considered as a manifestation of polymer δ - relaxation caused by the librational motion of monomer units of macromolecules.

4.2. Librational Movement and Low Temperature δ-Relaxation

As we saw above, the modes of molecular motion, with which α- and β-relaxations are associated in polymers, do not manifest themselves directly in the FIR range; their manifestation is more likely to be expected in the audio and microwave region. Nevertheless, FIR spectroscopy can still be an important source of information on the molecular mechanisms of these relaxations.

It is in the spectral range where the frequencies of resonance and relaxation processes are significantly approaching that information is contained on the molecular mobility that prepares α and β relaxation in polymers. We are talking about small-angle torsional vibrations (libration), which determine the absorption by the Poley mechanism in the range of 10 - 150 cm^{-1} (0.3 - 4.5 THz). The relationship between the spectral parameters of this absorption and the molecular characteristics of polymers was discussed above (see also Application 9.1).

As shown in (30, 31, 72, 73), based on the analysis of the absorption of Poley in the spectra of linear polymers, oligomers, and low molecular weight liquids, it is possible to estimate the potential barriers to the libration of monomer units and to establish the predominantly the intermolecular (cohesive) nature of these barriers, as well as to ensure their correspondence to the activation energies of δ - relaxation in polymers. On this basis, the molecular mechanism of the latter was unambiguously determined.

To assess the potential barriers to libration from FIR spectra, the authors used the Bro-Darmont model (see Section 4.1). Setting in the equation $U(\varphi) = U(0)\, sin^2(\pi\varphi/2\xi)$, $U(\varphi) = Q_{libr}$, we have at small libration angles $\nu_{libr} = \omega_0/2\pi c = [(Q_{libr}/2I)^{1/2}]/\pi c\varphi$. Note that the Bro-Darmont model forpolar liquids is in good agreement with experiment, and the potential barriers to the librational motion of molecules calculated with its help are close to those obtained by other methods [57, 113].

The analysis [72] proceeded from the FIR spectra of flexible-chain linear polymers with different structures: they either did not contain side groups in the chains, or contained small or massive side groups. The calculation was carried out under the assumption of the libration of the side group or when considering the monomer unit as a whole as a librator. The van der Waals volume of the librator was calculated, taken conventionally in the form of a ball, and its moment of inertia $I = 2MR^2_{eq}/5$ was determined, where M is the molecular mass of the librator, and R_{eq} is the equivalent radius. Based on theoretical concepts and calculations [111, 112], the libration angle of the monomer unit for the studied polymers was taken equal to 15^0. The libration angles of molecules of simple liquids are given in the literature [59, 113].

Table 2 shows the results of calculations for the FIR spectra of the values of Q_{libr}, which are compared with the cohesion energy E_{coh}: its values were taken from [88, 114]. In addition, the known values of the activation energy of the low-temperature δ-transition in polymers are given [115 - 119]. It can be seen that when performing the calculation for one monomeric unit of the polymer, satisfactory agreement is observed.

Table 2. Parameters of librational motion and low-temperature δ-relaxation in polymers and molecular liquids [72]

№	Material	Librational motion						$E_{coh}/3$ kJ/mole [27, 28]	Parameters of δ-relaxation	
		ν_{libr}, см$^{-1}$	φ, degrees	$I.10^{40}$, g·cm^2		Q libr, kJ/mole			Q_δ kJ/mole кДж/моль	T_δ, K
		[14-16, 19, 24]		Side group	Monomer unit or molecule	side group	Monomer unit or molecule			
1	CH_3CCl_3	27	35		715	-	9	9.5	-	35*
2	C_6H_6	75	25	-	340	-	14	11	-	44*
3	$CHCl_3$	35	30	-	525	-	7	8	-	30*
4	$(CH_3)_3CCl_3$	23	35	-	520	-	5	9	-	33*
5	C_6H_5Cl	44	30	-	570	-	13	14	-	55*
6	C_5H_5N	45	28	-	390	-	10	12	-	47*
7	$C_вH_5Br$	35	30	-	950	-	13	14,5	-	57*
8	CH_2Cl_2	78	48	-	50	-	9	9		33*
9	PE	100 [20]	20 [25]	-	76	-	2.7	2.6	2,2	40 (1 Гц) [2], 10*
10	PP	50	15 [25]	-	155	-	1.2	4.5	5	10-15 (1кГц) [3],18*
		100 [20]		-			4.8			
11	POE	105 [36]	15 [25]		140	-	44	3,2	-	13*
12	PVC	88 [13]	15[25]	52	210	2,5	5	6	5	18 (1 кГц) [29], 23*
13	PS	80 [13]	15[25]	345	590	7	11	12	9	45 (1 кГц) [8], 48*
14	PCS	45 [13]	15[25]	600	950	7	12	14	12	50 (1 Гц) [3], 51*
15	PMMA	95 [13]	15[25]	288	528	8	14	11	10	45 (300 Гц)[8],44*
16	PMA	83 [35]	15[25]	190	400	4	8	8	-	32*
17	PVA	86 [21]	15[25]	190	400	4	9	10	-	40*
18	PAN	127 [13]	15[25]	55	203	3	10	11	-	44*
19	PVF	88 [13]	15[25]	-	138	-	3	5	-	20*
20	PTFE	52 [33]	15[25]	-	372	-	3,2	2.5	-	10*
21	PPFS	40 [9]	15[25]	1210	1600	7	9	10-16	-	62*
22	PDMS	45 [23]	15[25]	-	500	-	3.5	5	5,5 [31]	23*
23	PMS	80	15[25]	345	815	10	16	15	10	35 (1 кГц) (30), 60*

Note. The values of Q_δ were obtained from the experimental values of T_δ from the Arrhenius relation $\upsilon - - 10^{13} exp\ (- Q/RT)$ at $\upsilon = 1$ Hz. Temperatures T * are predicted on the basis of the found correspondence $Q_\delta \approx E_{coh}/3$.

Consequently, low-temperature δ-relaxation in polymers corresponds to small-angle torsional vibrations of the monomer unit. Of course, in the case for macromolecules with massive side groups, it is difficult to separate the side group libration from the monomeric unit libration (see data on PS).

It is noteworthy that the found characteristic ratio of the values of Q_{libr} and E_{coh}, and namely: $Q_{libr} \approx E_{coh}/3$, indicating the determining contribution of intermolecular interactions to potential barriers to libration in comparison with the contribution of the barrier to internal rotation around the C – C bond. Obviously, this is related to the approximate coincidence of the position of the maximum of librational absorption in the FIR spectrum of the polymer and liquid, which is similar in structure to the monomer unit (for example, PS and C_6H_6 or PCS and C_6H_5Cl) (Table 2). It should be noted that the value $E_{coh}/3$ corresponds to the intermolecular potential barrier for the displacement of molecular units in the condensed system relative to the neighbors [120].

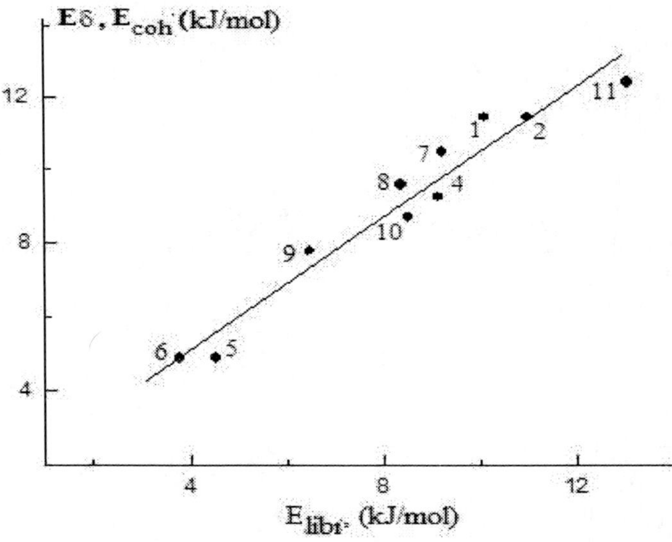

Figure 21. The relationship between the activation energy of the δ-transition for polymers E_δ (E_γ for liquids) and $E_{coh}/3$ with potential barriers to librational movement U_{liber}: 1-PS, 2-PCS, 3-PBMA, 4-PMMA, 5-PVC, 6-PVF, 7-PAN, 8-PVA, 9-$CHCl_3$, 10-CH_2Cl_2, 11- C_5H_6 Cl.

In polymers, this barrier corresponds to the activation energy of low-temperature δ-relaxation U_δ. The relationship between E_{libr}, $E_{coh}/3$ and U_δ is illustrated in Figure 21. It confirms the assignment of low-temperature δ - relaxation in polymers to small- angle high-frequency torsional vibrations of a molecular unit close in size to a monomer unit of a macromolecule.

Table 2 shows both the experimental temperatures of the δ - transition T_δ, and the calculated ones obtained from the Arrhenius ratio $\upsilon \approx 10^{13} exp(-Q/RT)$ at $\upsilon = 1$ Hz and $Q_{libr} = E_{coh}/3$. The found values are in the range of 20 - 70K, i.e., in the characteristic range of manifestation of excess (super-Debye) heat capacity and δ - relaxation in the spectra of mechanical losses. Showing the relationship between δ-relaxation and librational movement in macromolecules, it should be noted that a similar relationship can be traced for low-molar liquids. With regard to them, the possibility of obtaining from absorption by the Poley mechanism information about γ-relaxation (δ-relaxation in polymer terminology), caused by the librational motion of molecules, is discussed in [60, 93, 121].

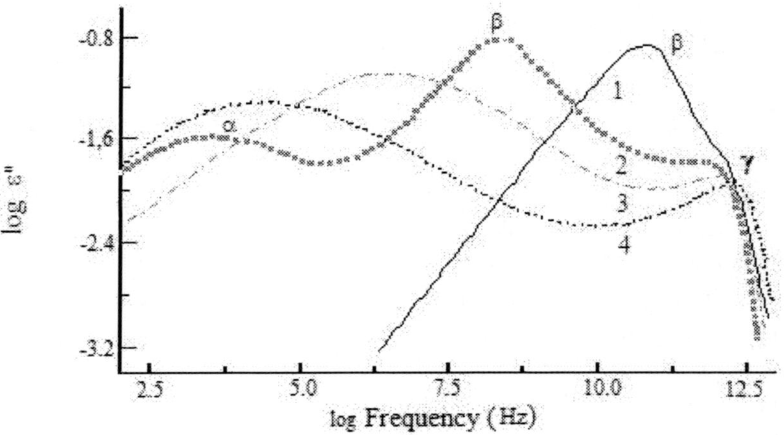

Figure 22. Dielectric loss spectrum of fluorobenzene at 293K (1), 110 K (2) and 143 K (3). From Reid and Evans, Ref. [60].

In Figure 22 shows the wide-range spectra of dielectric losses of a chlorobenzene solution in decalin at frequencies from 100 to $10^{12.5}$ Hz at various temperatures. At room temperature, the loss spectrum is mainly composed of a microwave peak at ~ 10^{11} Hz (~ 0.1 THz and ~ 3.3 cm^{-1}) with a poorly resolved shoulder at ~ $1.35 \cdot 10^{12}$ Hz, corresponding to the maximum $a(\nu)$ at 45 cm^{-1} in FIR spectra of chlorobenzene and PCS (see Figure 7d). With decreasing temperature, the peak of microwave losses, attributed by the authors of [60] to the manifestation of β - relaxation shifts to low frequencies. At 143 K, a peak corresponding to α-relaxation is already clearly visible in the spectrum, while the high-frequency part of the losses and the shoulder in the FIR range practically do not change their parameters. This high-frequency part of the dielectric losses in the FIR region is characterized by the authors as γ - process caused by the librational motion of the dipoles. γ- the process is universal and is a high-frequency precursor of α - and β - relaxations.

It is the γ - process (δ - relaxation in polymer terminology) that can be studied in more detail from the FIR spectra. The absorption coefficient $\alpha(\nu)$ is known to be related to the imaginary part of the complex dielectric constant $\varepsilon''(\nu)$ as follows: $\alpha(\nu) = 2\pi\nu \bullet \varepsilon''(\nu)/n(\nu)$, where $n(\nu)$ is the refractive index.

It should be added to what has been said that for polymers with a complicated structure or an increased length of the repeating unit, the made assignment of δ - relaxation may require correction. For example: for polyalkyl methacrylate's with long side alkyl radicals, the ratio Q_{libr} and *Ecoh/3* turned out to be fair, but only when calculating for a part of the monomer up to the 'oxygen hinge': i.e., excluding the alkyl radical. This part is the librator in this case. The movement of the radical determines the specific, polymeric γ-relaxation (see below).

Note that δ-relaxation in polymers is sometimes associated with rotation of the side methyl group. Indeed, in polymers containing this group, maxima of mechanical losses at low temperatures: for example, in PVA at 10K and Hz or at 10–20 K [144] and 300 Hz in PMMA [115]. As can be seen from Table 2, these temperatures lie below the region of manifestation of the relaxation condemned here, due to the libration movement of the monomer unit of the polymer molecule.

4.3. The Manifestation of the Boson Peak in the Low-Frequency IR Spectra of Amorphous Polymers

Before discussing the molecular mechanisms of other dielectric relaxations, which are manifested in the low-frequency IR spectra of amorphous polymers, let us consider a feature characteristic of these spectra, which is called the "boson peak" in the low-frequency Raman spectra (Figure 15, 23). The nature of BP is currently being widely discussed. A number of authors believe that the appearance of additional quasilocal vibrational modes is associated with the existence of medium-order and cohesive inhomogeneity in disordered media [122]. Others classify them as correlated vibrations: for example, in quartz - torsional vibrations in a chain of several SiO4 tetrahedra [123].

In [124], based on a comparative study of the IR and Raman spectra, these low-energy excitations in polymethyl methacrylate (PMMA) and its oligomers were presumably attributed to librational movement in regions of a polymer molecule consisting of several monomer units.

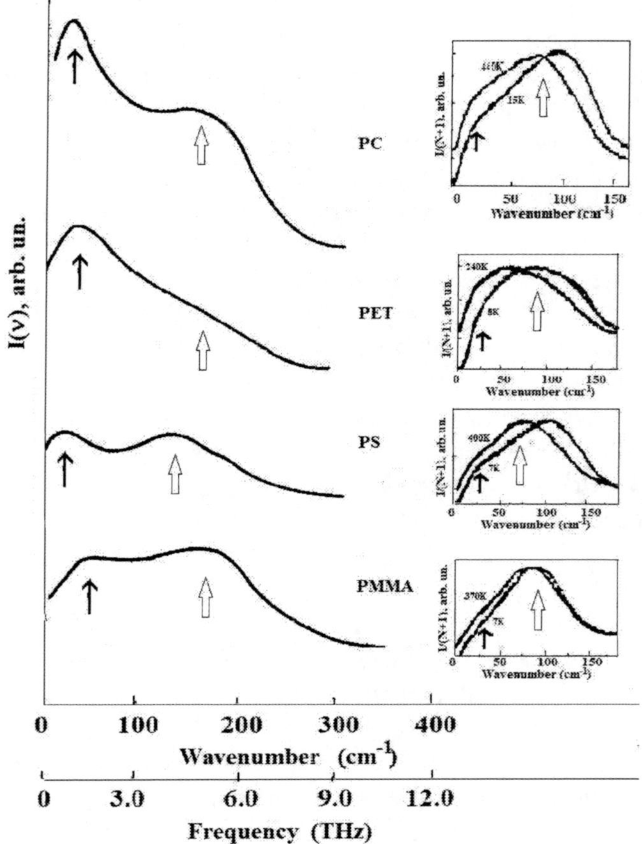

Figure 23. Initial *I (v)* and reduced (in the inserts) *I (v)/(N + 1)* Raman spectra of PMMA, PS, PET and PS [66, 127]. Arrows show: ↑ - boson peak; ⇑- librational absorption.

In Figure 24 shows the Raman and IR spectra of PMMA and its oligomers in the range 10 - 150 cm^{-1} in the coordinates $I_R = I(\nu)/(N(\nu) + 1)$ and $\alpha(\nu)/\nu$, respectively. They have a wide asymmetric band with a maximum at 80 - 90 cm^{-1} and an inflection at the frequency of the boson peak at 15 - 20 cm^{-1}, typical for low-frequency spectra of linear amorphous polymers.

Previous studies of low-frequency IR spectra of low-molecular-weight substances and polymers allow the band under consideration to be attributed to absorption due to libration (rotational vibrations) of monomer units in PMMA macromolecules, i. e., to absorption by the Poley mechanism [100].

The discussed band in the Raman spectrum of PMMA, which has a comparable contour and a close in frequency position of the maximum, should be attributed to the same motion [64].

A slight broadening of the band and a shift of its maximum towards low frequencies in the spectra of oligomers are apparently caused by an increase in the amplitude and anharmonicity of librational vibrations in low-molecular-weight analogs of PMMA.

The identity of the mechanisms of band formation in the low-frequency IR and Raman spectra of condensed media is confirmed by numerous data [125]. As discussed above, in one of the simplest models used to analyze absorption by the Poley mechanism [126], the librational motion of a molecule with the moment of inertia (I) occurs within the potential well formed by its closest environment, which has the form $U(\phi) = U_0 \sin^2 \pi\phi/2\xi$, with an angular frequency

$$\omega = 2\pi c\nu_{libr} \approx \pi/\xi\, (U_0/2I)^{1/2}, \qquad (7)$$

where U_0 is the depth of the potential well, ξ is its semi-angular aperture (width at a height equal to half of the full barrier), ϕ is the libration amplitude.

The model predicts the position of the maximum of the libration band in the low-frequency spectra of polar and non-polar liquids. The intermolecular barriers to librational movement found with its help correspond to those obtained by other methods [74].

A similar approach for polymers gives good agreement with experiment, provided that the librator is a monomer unit. In such calculations, it is assumed that the librator has the shape of a ball with a moment of inertia $I = 2MR^2/5$, where M is its molecular weight and R is its equivalent radius.

In the case of PMMA, an estimate of the librator size by $\nu_{libr} = 95$ cm^{-1} in expression (7), proceeding from the activation energy of the δ transition [72] and the presence of limited torsional vibrations with an amplitude of 10 - 15^0 [111] in macromolecules, gives the value R = 0.25 nm, which is close to the van der Waals radius of the monomer unit of this polymer.

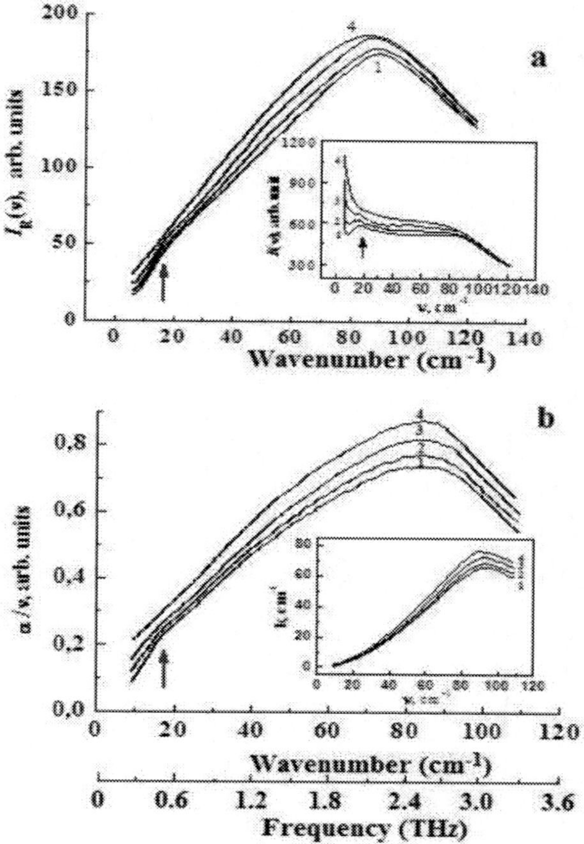

Figure 24. Reduced Raman- (a) and FIR-spectra (b) in the range of 10 - 130 cm^{-1} of the starting PMMA (1) and its oligomers with n = 9 (2), 7 (3) and 2 (4). The insets show their experimental Raman and FIR spectra at the same frequencies. Arrows indicate the position of the boson peak.

This result is an additional argument in favor of assigning the band with maximum at 95 cm^{-1} in the low-frequency Raman and FIR spectra of PMMA and its oligomers to the librational motion of the monomer unit of the macromolecule.

Let us turn to another feature of the low-frequency IR and Raman spectra of PMMA and its oligomers - an inflection in the low-frequency wing of the band under study. In the original (experimental) Raman spectrum of PMMA, the frequency of the boson peak maximum ν_{BP} is 16 – 18 cm^{-1} [127, 128]. After "reducing" the experimental spectrum to the coordinates $I(\nu)/(N(\nu)+1)$, the boson peak transforms into an inflection in the low-frequency wing of the band at the same frequency ν_{BP} (Figure 24a).

A comparison of the "reduced" Raman spectra with the IR spectra reduced to the $\alpha(\nu)/\nu$ coordinates (Figure 24b) shows that a similar inflection is also observed in the reconstructed PMMA spectra at the frequency of the boson peak $\nu_{BP} \sim 16$ - 20 cm^{-1}.

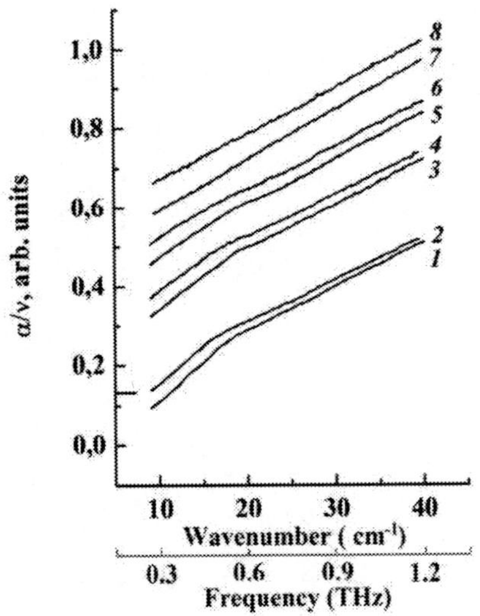

Figure 25. The reduced FIR spectra of PMMA (1, 2), oligomer with n = 9 (3, 4), oligomer with $n = 7$ (5, 6) and oligomer with $n = 2$ (7, 8). At T = 90 (1, 3, 5.7) and 293 K (2, 4, 6, 8) in the range of 10 - 40 cm^{-1}. The oligomer spectra are shifted along the ordinate axis.

In order to elucidate the molecular nature of the inflection on the low-frequency wing of the libration band, let us consider the reduced FIR and Raman spectra of PMMA and its oligomers (Figures 24 and 25). It follows from them that with an increase in the chain length, the inflection in the spectral curve shifts towards lower frequencies. The dependence of its position, ν_{BP}, on the oligomer chain length, expressed by the number of monomeric units *n*, is shown in Figure 26. The frequency of the main maximum of the band, caused by the libration of the monomer unit, also falls on the same dependence.

This indicates that the inflection on the low-frequency wing of the main band also characterizes librational motion, which, as the dependence of ν_{BP} on *n* shows, should be attributed to a motion in which several monomer units participate. The graph in Figure 26 allows one to estimate the maximum number of chain links involved in this correlated librational movement. In the case of PMMA and its oligomers, it does not exceed 5–7 monomers.

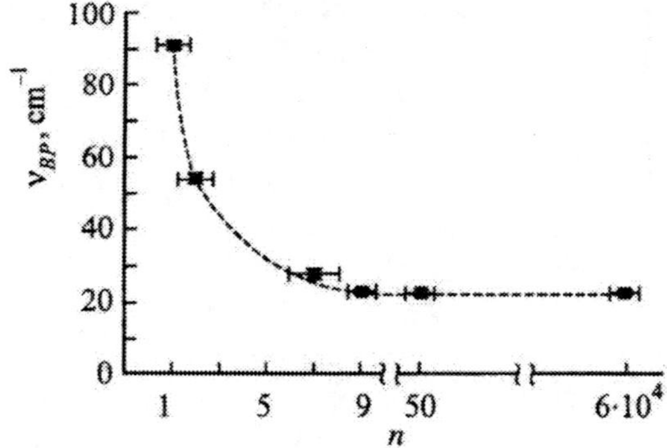

Figure 26. Dependence ν_{BP} on the degree of polymerization, expressed by the number of monomer units *n*.

The length *L* of such a section of the main chain can also be determined using the following expression [127]:

$$L \approx c_t\,(c \cdot \nu_{BP})^{-1}, \quad (8)$$

where *c* is the speed of light in vacuum, c_t is the "transverse" speed of sound (shear wave speed). For PMMA with $c_t \sim 1.42$ *km/s* [128]. Substituting this value and the value of the frequency of the boson peak $\nu_{BP} = 17$ cm^{-1} into (8), we find that $L \sim 3$ nm in PMMA. The obtained value corresponds to the length of a segment of the polymer chain of six monomer units and is close to the size of the statistical Kuhn segment for PMMA [129]. The state segment characterizes in this case the dynamic interconnection of neighboring units in the polymer chain.

It is noteworthy that the linear dimensions of the region of collective vibrational excitations in other glassy polymers are comparable with the dimensions of their statistical segments. Thus, from the analysis of the Raman spectra of glassy polystyrene (PS) and its oligomers [36, 130], it follows that the size of the indicated region here is 6–8 monomer units (the statistical segment of PS is equal to eight units [129]). For polybutadiene [130] and polypropylene glycol [131], the region of collective vibrational excitations includes 11 and 7 monomer units, respectively, which is also close to the values of the statistical segments for these polymers.

If we substitute the value of the frequency of the maximum of the libration band ν_{libr} = 95 cm^{-1} into expression (8), then the found value of the length of the segment of the polymer molecule turns out to be equal to 0.5 nm, i.e., the length of the monomer unit in PMMA.

Thus, a broad band in the low-frequency IR and Raman spectra of PMMA and its oligomers with a maximum at 95 cm^{-1} characterizes the torsional vibrations of the polymer monomer unit, while the inflection on its low-frequency wing at 16-18 cm^{-1} is due to correlated torsional vibrations involving several monomer units.

It is essential that the frequency ν_{BP} found from the low-frequency IR Raman spectra the linear size of the region of correlated librational excitations corresponds to the length of the statistical segment - the minimum fragment of the polymer chain required for the rotational-isomeric transition to occur in it. The movement of segments of this length, occurring in the places of the least dense packing of macromolecules, underlies β-relaxation [129, 132]. It is analogous to the Johari – Goldstein relaxation [93] in low molecular weight vitrified liquids, which consists in the reorientation of the molecule under the action of thermal fluctuations. In this case, the molecule overcomes the potential barrier U_0 formed by its nearest environment.

The β-transition is preceded by a universal γ-process [60]: small-angle torsional vibrations - libration of molecules, which, as already noted, causes the appearance of the Poely band in the low-frequency IR and Raman spectra of condensed media at a frequency ν_{libr}.

That is, from the low-frequency IR and Raman spectra, it is possible to directly determine both the parameters of the librational motion, which prepares the β-relaxation, and the sizes of the regions of the polymer chain involved in this movement during the β-transition.

Let us turn again to Figure 24. It can be seen from it that the intensity of the low-frequency wing of the libration band (in the frequency range below

ν_{BP}) increases with increasing temperature. A similar effect was previously observed in the Raman spectra of glassy solids [127, 133]. According to these studies, the increase in intensity below ν_{BP} is due to the contribution of the β-process increasing with temperature. Consequently, the inflection frequency in the low-frequency IR and Raman spectra of PMMA corresponds, in addition, to the boundary of the transition from the resonant type of absorption to the relaxation one, when the number of monomers involved in the librational motion turns out to be sufficient for the conformational rearrangement in the polymer chain.

Thus, the analysis of low-frequency IR and Raman spectra of PMMA and its oligomers showed the following.

1. The studied band - the Poley band - refers to the librations of the monomer unit, causing the universal γ-process (δ-transition in polymer terminology).
2. The maximum length of the correlation region of the macromolecule found from the inflection frequency on the low-frequency wing of this band ν_{BP} is close to the size of the statistical segment, the motion of which underlies the relaxation β-transition.
3. Frequencies below the frequency of the bosonic peak ν_{BP} correspond to the relaxation mobility of macromolecules, which is realized when the number of monomers involved in librational movement becomes sufficient for a conformational transition in the polymer chain.

Let us present one more work in which the dynamics of the boson peak was studied using both Raman and far IR spectroscopy in the terahertz range [134]. The spectra of a natural polymer of starch, which includes two polymers: amylose and amylopectin were obtained in a wide temperature range from 33 to 300 K. In the frequency range 0.25 - 2.25 THz, the spectra were a superposition of two contributions: the boson peak (the contribution of the excess density of vibrational states (VDOS)) and relaxation (contribution of fast relaxations). The latter grew with temperature and at 180K completely hid the boson peak. In Figure 27 shows the Raman and terahertz spectra of starch. Raman spectra in coordinates $I(\nu)$ (initial) and $I(\nu)/(N+1)$ (reduced); terahertz in the α/ν^2 coordinates (since, according to the authors, in this case, the excess density of vibrational states appears as a peak in the VDOS spectrum) and in the α/ν coordinates (reduced). In the initial Raman spectra, a boson peak is observed at 1.1 THz and a temperature of 33 K, in terahertz spectra at 0.99 THz. In the reduced Raman and terahertz IR spectra, the spectral curve has an inflection at the same frequencies.

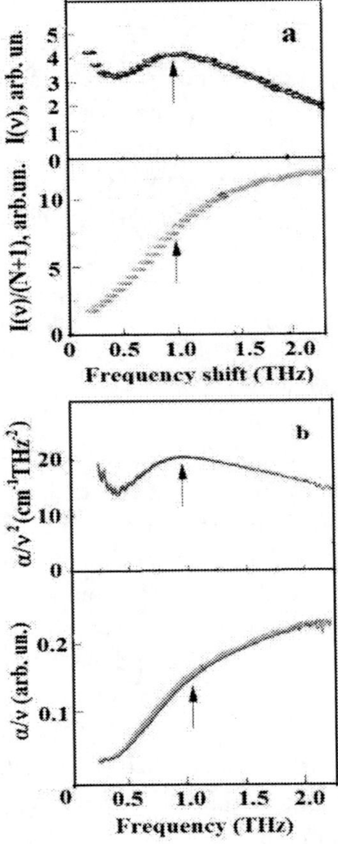

Figure 27. Experimental *I (ν)* and reduced *I (ν)/(N + 1)* Raman spectra (a) of starch and its terahertz IR spectra (b) in coordinates α/ν^2 and α/ν. The arrow shows the position of the boson peak. Ref. [134].

Comparison of the study results with the results of a similar study of the spectra of glucose, which is close in its chemical structure to the starch monomer unit, showed that both the frequency of the boson peak and the absorption coefficient in the case of starch are significantly lower. This discrepancy is associated not only with the lower concentration of hydroxyl groups in starch. A more significant reason, according to the authors, is the large mean order correlation length in the polymer as compared to the monomer, which is consistent with the conclusions of [135]. Note that this feature of the manifestation of the boson peak in the FIR and Raman spectra of polymers in comparison with the higher-frequency position of the boson peak in the spectra of vitrified liquids leads to the fact that it is often referred to as librational absorption [121] (see also Application 9.2).

4.4. Manifestation of Fast Relaxations in the Terahertz Spectra of Amorphous Polymers

In amorphous polymers, a part of the terahertz loss spectra at frequencies below the beginning of the manifestation of the boson peak or, more precisely, the vibrational density of states (VDOS), i.e., below ~ 0.5 THz, is very close to zero, but in fact, as can be seen from Figure 14, 15 is not zero. Relaxation losses at these frequencies are due to local dynamics described by the coupled mode model (MCT) [103], as the vibrational ("rattling") motion of a molecule in a cell formed by its immediate environment. This process is sometimes also called fast β-relaxation, although this term is not correct in the sense that in reality no separate relaxation process can be associated with this movement, because it covers a wide range of frequencies and exhibits only a small frequency dependence. Hence, the name of the region ~ 10^8 - 10^{11} Hz as the nearly constant loss region (NCL).

The nature of fast molecular processes observed in low-frequency spectra is currently being actively discussed in the literature, including on the basis of data obtained by long-wavelength and terahertz IR spectroscopy, which makes it possible to specify their mechanisms [136, 137]. In the model of coupled modes of the theory of glass transition (MCT), these processes are attributed to the manifestation of β - relaxation preceding structural relaxation [138]. The contribution of elementary processes such as local conformational reorientations in the main chain and side groups is also considered [139]. This ambiguity in the assignment of fast processes in polymers is explained by the fact that it is often carried out using phenomenological concepts based on such averaged characteristics of a substance as free volume, coefficients of viscosity, diffusion, etc.

Molecular kinetic theories based on the concept of the genetic relationship between vibrational and relaxation forms of motion [132] make it possible to more confidently concretize the mechanisms of molecular movements responsible for the picosecond dynamics of molecules. An example of such an approach can be found in Ref. [133], in which it was found that the relaxation component at the quasi-elastic scattering frequencies in the Raman spectra of polystyrene and polycarbonate measured in the range 3 - 1000 GHz at temperatures below the glass transition temperature (T_g) is due to the temperature activated transitions through a barrier with a height of ~ 500 K [140].

The ability to determine the value of the activation energy of fast rearrangements in polymethyl methacrylate (PMMA), based on the change in spectral parameters of low-frequency absorption with temperature was also used in [106] to elucidate the nature of the relaxation contribution to the FIR spectra of this polymer.

The FIR spectra of PMMA at different temperatures are shown in Figure 28 in the coordinates $\alpha(\nu)/\nu$.

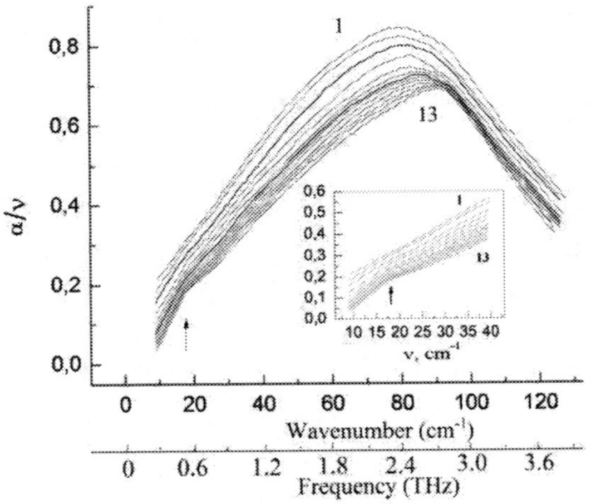

Figure 28. Reduced FIR spectra of PMMA in the range of 10 - 130 cm^{-1} at temperatures T, K: 1- 413, 2 - 403, 3 - 378, 4 - 333, 5 - 293, 6 - 273, 7 - 253, 8 - 233, 9 - 1939, 10 - 173, 11 - 153, 12 – 123 and 13 – 90K. The inset shows the initial sections of the curves on an enlarged scale, arrows mark the position of the boson peak in the spectra.

The representation of the FIR spectra in such coordinates allows, as noted above, to compare them with the Raman spectra presented in the coordinates $I(\nu)/(N+1)$, where I (ν) is the measured scattering intensity, and N + is the Bose-factor and also with a spectrum of dielectric losses in the coordinates ε'' *vs* ν (cm^{-1}).

Like the reduced Raman spectra of PMMA recorded in the range of 5 - 150 cm^{-1} for different temperatures [141], the reduced FIR spectra of PMMA have here an intense band at ~ 90 cm^{-1} with an inflection at frequencies close to the position of the boson peak in the Raman spectra.

Figure 28 that as the temperature rises, the absorption band at ~ 90 cm^{-1} broadens somewhat and its intensity grows, and the maximum shifts by 3-5 cm^{-1} to low frequencies. Of particular interest is the increase in the intensity of its length of the low-frequency wing at frequencies below the BP, indicating an increase in the relaxation component of the spectrum.

The dynamics of the growth of the intensity of low-frequency wing of the band is illustrated in Figure 29, which shows the difference spectra obtained by subtracting from them the PMMA spectrum at 90 K, which, as follows from Ref. [141], is mainly a purely vibrational contribution.

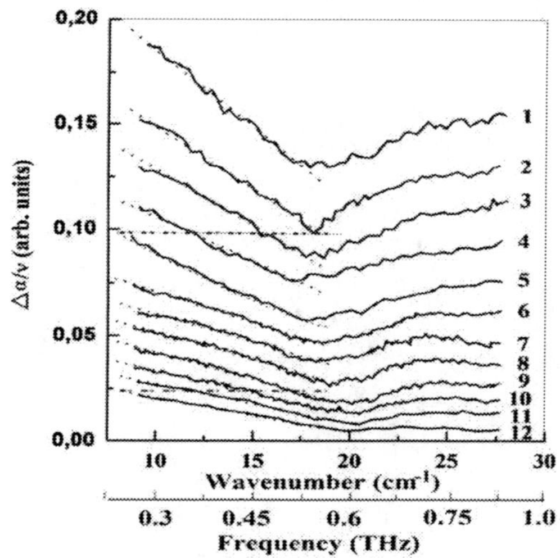

Figure 29. Spectra of the difference between the FIR spectra of PMMA at temperatures from 420 to 123K and the spectrum of PMMA at 90 K. The designations are the same as in Figure 28.

As can be seen from the plot of the dependence of the intensity of the long-wavelength wing of the difference spectrum (DS) on temperature (Figure 30), the increase in absorption in the range 8 - 20 cm^{-1} is not monotonic. It clearly shows two stages: "low-temperature" from 90 to ~ 260K and sharper "high-temperature" from ~ 260 to 420K. The temperature at which the conformational mobility in PMMA begins is close to 260 K

[129], which suggests that a sharp increase in the DS intensity is due to just such a process.

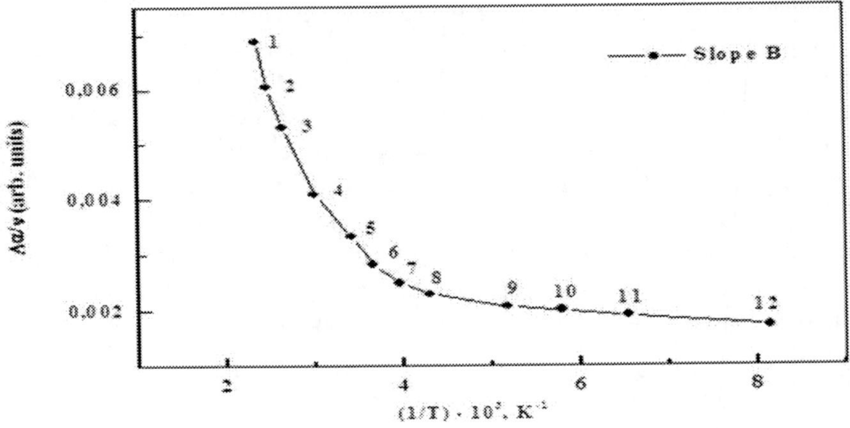

Figure 30. Temperature dependence of the intensity of the FIR spectra of PMMA in the range 8 - 20 cm^{-1}. The designations are the same as in Figure 28.

To confirm this assumption, let us estimate the height of the potential barrier of the "high-temperature" relaxation process using the available experimental data on its temperature shift. For example, as is done in the method of dynamic mechanical analysis, DSC or dielectric measurements, when the activation energy of the relaxation process is determined from the graph of the linear dependence of the frequency shift of the relaxation peak f with temperature $ln\, f\, (T^{-1})$ and the formula $E = - R\, d\, (ln\, f)/d\, (1/T)$, where R is the universal gas constant, and $f = cv$, where c is the speed of light in vacuum.

Since at high frequencies the relaxation peak shifts parallel to itself upon variation of T, keeping the amplitude and half-width [142], this dependence can be plotted not only from the position of the maximum, but also from the displacement of the relaxation process wing. In our case, the peak of relaxation losses does not fall into the investigated wavelength interval, therefore, its shift was determined from the DS shift with a variation of *T*. In Figure 29, the change in the position of the DS at $\Delta\alpha/\nu$ "high-temperature" and at $\Delta\alpha/\nu$ "low-temperature" relaxation processes are shown by horizontal dashed lines.

The dependences ln f (1/T) obtained in this way are shown in Figure 31 (the insets illustrate the shift of the "peaks" of the distinguished relaxation processes). From the slope of these dependences (B) and the formula $E = -8.3\ B$ (J/mol), the activation energies of the "low-temperature" and "high-temperature" processes were determined, which amounted to ~ 2 and 8 kJ/mol, respectively.

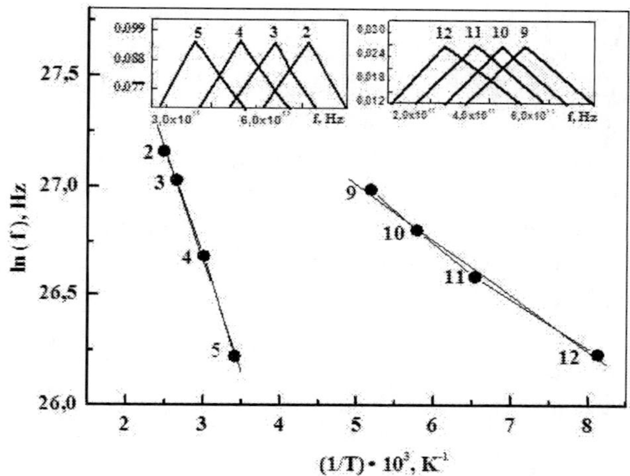

Figure 31. Dependences of the frequency position of "high-temperature" (2-5) and "low-temperature" (9-12) relaxation losses from temperature. The insets schematically illustrate the shift of the peak of these losses with temperature. Legend is the same as in Figures 28, 29 and 30.

The activation energy of the first of them is close to the activation energy of fast relaxation in polyvinyl acetate (1.9 kJ/mol), referred to 180^0 rotations of the COC group [143].

The barrier of the second fast process corresponds to the energy difference of the rotational isomers in PMMA, and the fact that it begins to manifest itself at the temperature of the appearance of conformational mobility in PMMA allows us to consider it a high-frequency analog of β-relaxation in this polymer.

4.5. Skeletal Torsional Vibrations and the General Nature of α and β-Relaxations

The study of β - relaxation process in polymers at T < Tg - temperature close to the glass transition temperature: is the subject of great scientific and technological interest. Until now, there has been no unambiguous assignment β-relaxation, which could be explained either by the absence of its general mechanism for polymers of different structures, or by insufficient understanding of this phenomenon due to the lack of direct experimental data.

Recently, new results have been obtained using dielectric, FIR and Raman spectroscopy, DSC and other techniques, which allowed formulating the general concept of the β - transition and obtaining experimental proof of this concept. According to [145-147], β-relaxation consists in the rotational (reorientation) movement of a chain fragment close in magnitude to the correlation portion - the statistical Kuhn segment. The barriers opposing this movement are mainly intermolecular, but a single-barrier trans-gosh transition is also involved in the process.

As shown, the special significance of β - relaxation is that. that the inherent acts of motion, carried out quasi-independently or collectively, actually control the segmental mobility at all temperatures $T \geq T\beta$, including the α-glass transition, high-elasticity regions and polymer flow, and, as a result, the physical properties and chemical processes in polymers controlled by segmental dynamics [146 -148]. Experimental facts related to the β - transition are discussed in the work of Bershtein and Egorov [129]. Below are a number of results related to the elucidation of the nature of β-relaxation in polymers using FIR spectroscopy.

In [77], the effect on low-frequency skeletal oscillations of the introduction of different, dosed concentrations of chemical crosslinks into a linear polymer was studied. The experiments were carried out on copolymers of styrene (S) with divinylbenzene (S-DVB) and methyl methacrylate (MMA) with dimethacrylateethylene glycol (MMA-DMEG) obtained by thermal polymerization. In these copolymers, the concentration of the second component, i.e., rigid links in the case of DV and relatively flexible in DMEG containing oxygen hinges varied from to 80 - 100%.

The difference between the bands of skeletal vibrations at 245 and 225 cm^{-1} in the spectrum of RS and PMMA, on the one hand, and by them in the spectra of copolymers, consists in a decrease in the intensity of these bands (Figures 32 and 33). This decrease becomes noticeable, when the concentration of the second component is of the order of 5-10 mol%. The induced decrease in the intensity of low-frequency skeletal oscillations is obviously caused by both a decrease in the amplitude of these oscillations and a decrease in the number of oscillating units (absorbing centers).

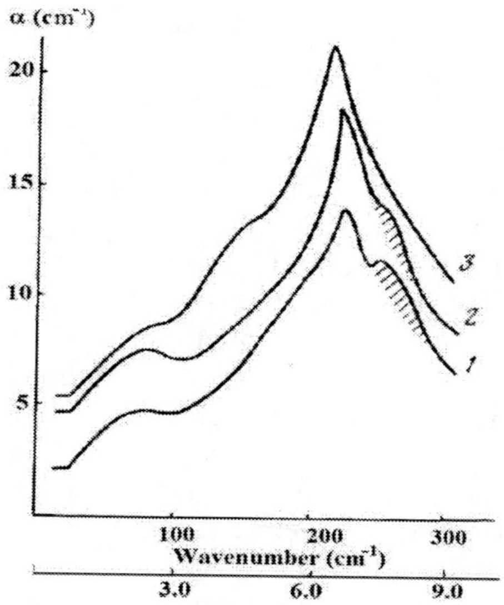

Figure 32. Far infrared spectra of PS and S-DVB copolymers: Curves 1, PS; 2, 3 mole % and 3,25 mole % DVB. Spectra 2 and 3 are shifted along the ordinate.

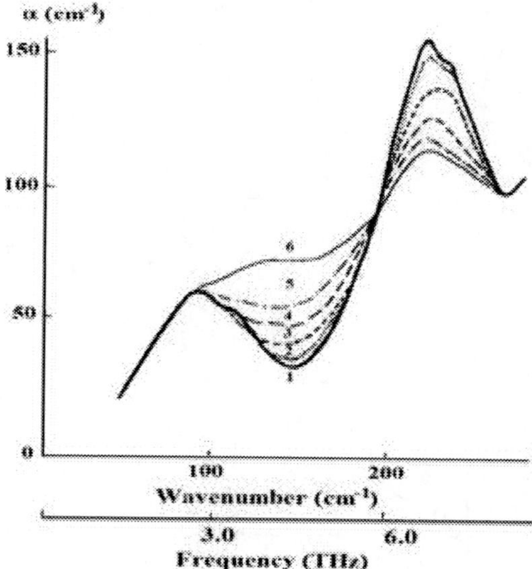

Figure 33. Fare infrared spectra of PMMA, Curve and copolymers of MMA with DMEG at various concentrations of the latter: 2,4 mole %; 3,5 mole %; 5,33 mole % and 6,100 mole %, (Curves 2-5).

In addition, as the crosslinking progresses, an increase in absorption is also observed in the frequency range 120-150 cm^{-1}, most significantly at SDRD $\geq$ 25 mol% and SDMEG = 33 and 100 mol%. The analysis showed that this absorption cannot be attributed to the vibrations formed during sewing of cross bridges or to mesh defects. At the same time, as already noted in Chapter 2, according to calculations in the region of 100-120 cm^{-1}, absorption corresponding to torsional vibrations in the main chain should appear in caro-chain polymers. It can be assumed that the crosslinks, limiting the correlated skeletal torsional oscillations that appear at higher frequencies, allow only limited torsional oscillations with smaller displacement amplitude and smaller chain fragments. This issue is discussed in section 5.5.

In Figure 34 shows the experimental dependence of the relative change in the intensity of the bands of torsional skeletal oscillations A_{sk} for the considered copolymers from the average distance (the number of monomer units N) between crosslinks. Here, the values N = 0.5 and 0.1 correspond to the most strongly crosslinked networks with a concentration of C_{DVB} = 82 mol% and C_{DMEG} = 100 mol%. The value A_{sk} = A_{max} corresponds to the difference between the intensity of the skeletal vibration bands for a linear polymer and its intensity for a tightly crosslinked copolymer. A_{sk} <A_{max} characterize the decrease in the intensity of these bands (at 245 and 225cm^{-1}) with different degrees of chain crosslinking.

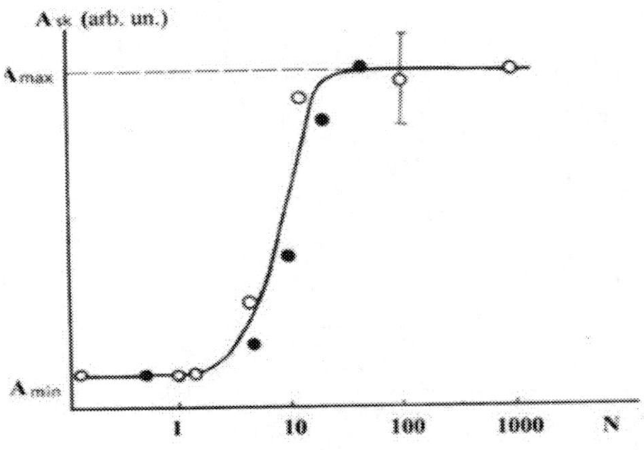

Figure 34. Intensity (A_{sk}) of the 245 and 225 cm^{-1} bands as a function of the mean spacing between chemical cross-links expressed in the number of monomeric units (N). Open circles, copolymers of S-DVB; filled circles, copolymers of MMA-DMEG,

Figure 34 shows that the A*sk* (N) dependence has a transition section from the intensities characteristic of PS and PMMA at N ≤ 15-20 to their largest drop at N ≥ 5. The considerable width of this section is obviously due to the dispersion of cross-link spacings. The center of the transition section falls on the values of N ≈ 8-10 monomer units. This value is close to the size of the correlation section, characterizing the equilibrium rigidity, i.e., Kuhn segment, equal for PS and PMMA, respectively: S = 6 and 8 monomeric units of the main chain [149]. Consequently, the intensity of skeletal torsional vibrations changes most sharply when the distance between the stitches, which have a braking effect on the movement, becomes commensurate with the length of the Kuhn segment or less than it.

The relationship between the parameters of the IR absorption bands and the length of the polymer chain was considered in the calculations of the frequencies and intensities of the bands performed by Gribov [150] and Zerbi [151], which showed that the most significant changes in the spectra should be expected in the series of oligomers with a length of several monomeric units, further growth the chain changes the spectrum slightly.

The regions of conformational regularity [11] and the size of the vibrational segment in polymers as applied to oligomers PMMA [12] and PS (152) were estimated experimentally from IR spectra. However, the determination of the value of the vibrational segment from the spectra in the mid-IR region in the bands characterizing only intramolecular vibrations is still not entirely correct.

Below, we consider the authors' data on the estimation of the unit of motion in torsional skeletal vibrations from the FIR spectra by using as samples, along with polymers, oligomers with different average degrees of polymerization (*n*). The experiments were carried out on the series of poly-α-methylstyrene (PMS) with *n* = 4, 7, 10, 14 and 2 × 10^2; MMA oligomers with *n* = 2, 7, 9, and 50 units and PC oligocarbonates with n = 2, 3, 4.6, 7, and 1.3 × 10^3.

As can be seen from Figure 35, the most significant changes in the spectra of MMA oligomers at ν> 130 cm-1 take place on going to the lowest molecular weight oligomers c. n = 2 or 7. Beginning with n = 9, in the frequency range 130-280 cm-1, the shape of the spectrum is almost stabilized and becomes close to polymer. The changes concern, first of all, the band of torsional skeletal vibrations at 245 cm^{-1}, not only sharply decreasing in intensity, but also acquiring a complex structure, as well as the appearance of a new band with a maximum of 130 - 160 cm^{-1}, appearing in approximately the same spectral range as in the crosslinking of linear polymers. Recall that the value of the Kuhn segment in PMMA is S = 6.

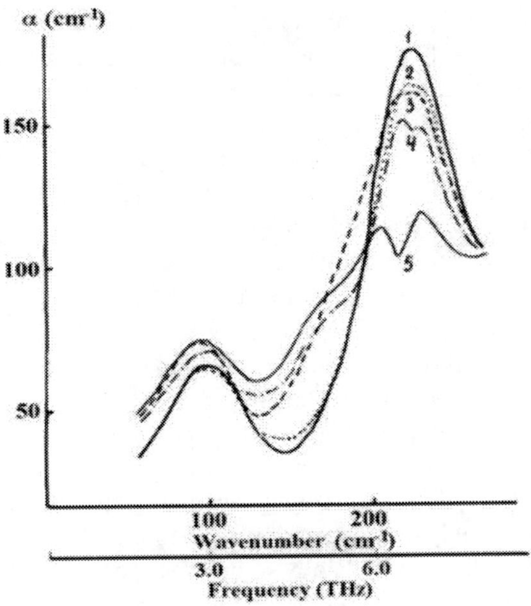

Figure 35. Far infrared spectra of PMMA (1) and oligomethacrylates with n = 50 (2), n = 9 (3), n = 7 (4) and n = 2 (5) at the temperature of liquid nitrogen, Ref. [135].

In the case of oligocarbonates, the FIR spectrum differed most significantly only for the dimer, but, already, starting with the trimer, the spectrum in the region of 25 - 300 cm^{-1} practically coincides with the spectrum of PC. Similarly, in the PMS series, the FIR spectra of oligomers were close to the polymer spectrum, starting from $n = 10$ (Figure 36).

Figure 36 shows the experimental dependences of the change in the intensity of the band of skeletal vibrations on the average length of the oligomer molecule, expressed by the number of monomer units n for PS and PMS. It is seen that the dependence A_{sk} (n) has an area of sharp change in intensity with increasing n, after which it remains practically unchanged. The shaded stripes in this figure correspond to the sizes of the statistical segments in the PMS and PS, equal to 8 and 2 monomer units, respectively [86]. Consequently, the changes in the absorption intensities of the bands of torsional skeletal vibrations in the oligomeric-polymer series are almost completed after the chain length exceeds the size of the statistical segment.

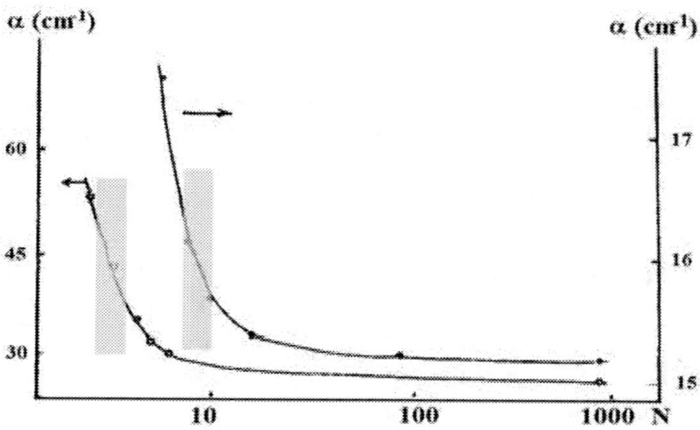

Figure 36. Dependence of the intensity of low-frequency skeletal band in the far infrared spectra of PMC (•) and PC (○) on the degree of polymerization (N). The shaded areas correspond to the Kuhn segment value, Ref. [82].

Important information for elucidating the nature of β-relaxation in flexible-chain polymers was obtained in [153] when evaluating potential barriers Q_{sk} and the scale of units of motion during torsional skeletal oscillations from the FIR spectra. In doing so, we proceeded from the following:

In the case of the simplest hydrocarbon molecules, for example, substituted ethane, the potential barrier of hindered internal rotation can be represented as (100): $Q\ (\varphi) = Q_0\ [(1 - Cos\ 3\varphi) + Q\ (r_{ik})]/2,$ where Q_0 is the torsional (orientational) barrier of hindered rotation around a covalent bond in the absence of nonbonding interactions, and $Q\ (r_{ik})$ is a steric barrier to rotation created by nonbonding interactions of substituents separated by a distance r*ik* and adjacent to this bond. In the simplest case of ethane, when the steric potential can be neglected due to the small van der Waals radius of the hydrogen atom, the complete barrier to hindered rotation around the C - C bond is determined mainly by the value $Q\ (\varphi) = Q_0\ [(1 - Cos\ 3\varphi)]/2$, where $Q_0 = Q_{C\text{-}C} \approx$ 15-16 kJ/mol [112, 144].

In this case, assuming a simple oscillatory motion, we have the frequency of the first torsional transition in the form $\nu = [\pi c\ (f/I)^{1/2}]/2$, where $f = \partial Q^2\ (\varphi)/\partial\varphi^2$ is the force constant, and I is moment of inertia of the molecule. The frequencies of torsional vibrations for substituted ethane, obtained on the basis of this simplest ethane potential, are in the range 120-280 cm^{-1}, which is in agreement with experiment (147).

For carbon-chain polymers, a calculation proceeding from $Q_{sk} = Q_{calc} = Q_{C\text{-}C}$ predicts the presence of bands of torsional vibrations of the carbon-carbon chain in the range 100-200 cm^{-1} [132], in particular, for PE $\nu_{calc} = 140$ cm^{-1} [5]. However, experimentally in the FIR spectrum of PE, the absorption due to torsion skeletal vibrations appears as a wide band centered at 200 cm^{-1}, and in the spectra of other linear polymers at frequencies $\nu_{sk} = 170\text{-}260$ cm^{-1} (Table 3).

This discrepancy between the experiment and the calculation can most naturally be explained by the fact that it does not take into account the total steric potential - the term $Q\ (r_{ik})$ in the expression for $Q\ (\varphi)$, i.e., indispensable involvement in the torsional movement of adjacent links of the correlation section of the chain adjacent to the axis of rotation. Especially when turning through large angles, when the torsional vibrations of adjacent monomer units can no longer be considered independent. In this case, the potential barrier to turning movement in the chain will be determined not only (and even not so much) by the value of Q_0 as by the interactions of covalently unbound atoms of the entire correlation section involved in the motion, outside of which there is no chain 'perturbation' by this motion, i.e., $Q_{sk} = Q_{C\text{-}C} + \sum Q\ (r_{ik})$, where summation over all links in the correlation section. An indication of the important role of non-bonded: cohesive interactions in this case is the experimentally observed proportionality of the frequency of torsional skeletal vibrations to the value of E_{coh} (Figure 5).

Table 3. Comparison of the parameters of low-frequency skeletal vibrations and β - relaxation parameters

No	Polymers	ν_{sk} (cm^{-1})	Q_{sk}	Q_{β}	N	S^a
			($kJ\ mol^{-1}$)		(monomer units)	
1	PE	208	32	36	7	8
2	PP	169	62	50	9	9
3	PVC	195	44	62	6	12
4	PS	245	86	100	7	8
5	PCS	255	100	90	6	8
6	PMMA	225	77	85	5	6
7	PMA	200	49	42	4	6
8	PDMS	190	48	40	8	5
9	PMS	245	80	110	7	8
10	PAN	250	90	110	7	9
11	PVA	232	67	58	6	7

[a] Number of monomer units in the statistical segment.

Since the frequency of torsional vibrations ν_{sk} is known from the experiment, it is possible, by comparing its value with the calculated frequency Q_{calc}, to estimate the value of the barrier Q_{sk} from ratio $(\nu_{sk}/\nu_{calc})^{1/2} = Qsk/Q_{calc}$. Taking Q $(r_{ik}) = E_{coh}/3$, i.e., equal to the barrier for the torsional vibrations of the monomer unit (the possibility of estimating Q_{libr} from the FIR spectrum of the polymer and its approximate equality $E_{coh}/3$ was discussed above) and $\sum Q(r_{ik}) = = NQ_{libr}$, we obtain that $Q_{sk} = Q_{C-C} + NQ_{libr}$, and the number of into the torsion-oscillatory motion of the chain section, i.e., the value of the torsion-vibrational segment $N = (Q_{sk} - Q_{C-C})/Q_{lib.}$

The experimental values of the frequencies νsc for 11 hycochain polymers of various structures are given in Table 3 together with the activation energy Q_β and the temperature β-transition.

T_β, obtained using DSC and conventional relaxation methods [147, 148]. In addition, the table shows the values of the Kuhn segment for these polymers. It can be seen that in almost all cases the barriers Q_{sk} are close, with an accuracy of 20%, to the activation energies Q_β (Figure 37), and the magnitude of the torsion-vibrational segment N to the magnitude of the statistical segment S. Thus, the results of [153] are in agreement with the above notion of the β-transition mechanism, which is common for flexible-chain polymers of different structures.

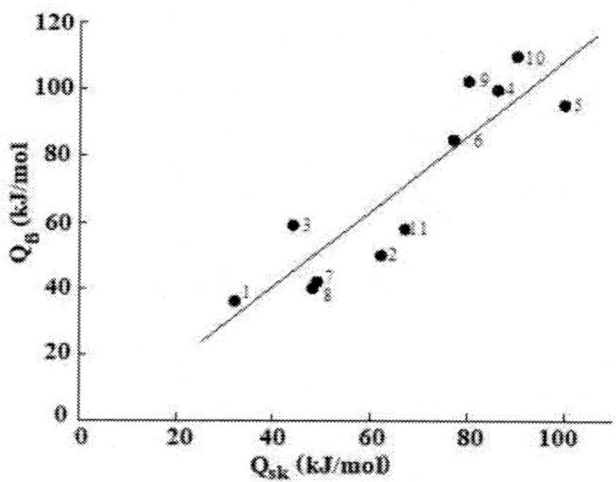

Figure 37. Relationship between the activation energy of β-transition and the potential barriers of low- frequency skeletal vibrations. Points 1-11 as in Table 3, Ref. [120].

4.6. The Onset of Conformational Mobility in Chains and Characteristic Temperatures in Amorphous Polymers

The considered model of the *β*-relaxation act in flexible-chain polymers provides for the participation in it, in addition to the torsional motion of the Kuhn segment, of a single-barrier trans ↔ gauche transition. The possibility of a conformational transition at $T < Tg$ was shown in [154, 155] theoretically and using mid-IR spectroscopy from the dynamics of hydrogen bonds in glassy polymers. This also agrees with the fact of the onset of physical aging in them, which is certainly associated with molecular rearrangements, precisely from the *β*-relaxation region [129, 148].

The found relationship between skeletal torsional vibrations in chains and the value of the Kuhn segment (thermodynamic rigidity of the chain) made it possible to expect the manifestation of differences in the conformational structure of the polymer in the bands of low-frequency skeletal vibrations. In addition, the conformational sensitivity of the bands of low-frequency skeletal vibrations was noted earlier in [40, 156].

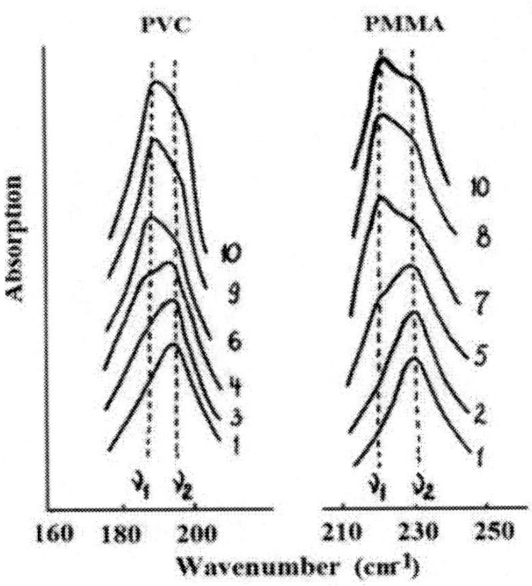

Figure 38. Changes in the band shape of the skeletal torsional vibration in the far infrared spectra of PVC and PMMA in dependence on the temperature: (1) 85 K, (2) t45 K, (3) 205 K, (4) 225 K, (5) 250 K, (6) 270 K, (7) 273 K, (8) 285 K, (9) 298 K, (10) 315 K, Ref. [33].

And in [33], as a result of measuring the temperature dependences of the spectra of PVC, PMMA, and PAN in the range from 90 to 400 K, direct information about the onset of defrosting conformational flexibility in these polymers.

In Figure 38 shows fragments of the FIR spectra of PVC and PMMA in the region of the bands torsional skeletal vibrations and their change with temperature. The contours of the bands include two components with a maximum of 185 and 195 cm^{-1} for PVC and 220 and 230 cm^{-1} for PMMA at room temperature. The appearance of these doublets (PAN at 240 and 257 cm^{-1}) was due to the presence of two rotational isomers in the chain. Indeed, according to [112], the energy difference between isomeric (trans - gauche) states $\Delta E = E_g - E_t = [RT_1 \cdot T_2/(T_1 - T_2)] \cdot ln\ [\alpha\ (\nu_2, T_2) \cdot \alpha\ (\nu_1, T_1)/\alpha\ (\nu_1, T_2) \cdot \alpha\ (\nu_2, T_1)]$, where $\alpha\ (\nu_1, T_1, T_2)$ and $\alpha\ (\nu_2, T_1, T_2)$ are the intensities of the muzzle bands measured at two temperatures ($T1 > T2$). The estimates made in [33] gave the values ΔE = 5-10 kJ/mol, which are close to those expected for the studied polymers [157].

In Figure 39 shows the dependence of the parameter $\alpha\ (\nu_1)/\alpha\ (\nu_2)$. As can be seen, a qualitatively similar picture is observed for all studied polymers. Up to certain temperatures, the parameter α (ν1)/α (ν2) remains unchanged, and then begins to grow due to an increase in the intensity of the low-frequency component. In the case of PVC, this growth starts at 220K, for PMMA at 250K and for PAN at 330K. These temperatures are significantly below T_g, equal to 350, 380 and 390 K at Hz, respectively. The found temperatures (T^* according to the terminology [33]) are rather close to the β-transition temperature equal to 230-250K for PVC, 280-290K for PMMA, and 330-340K for PAN at 1Hz [129, 147].

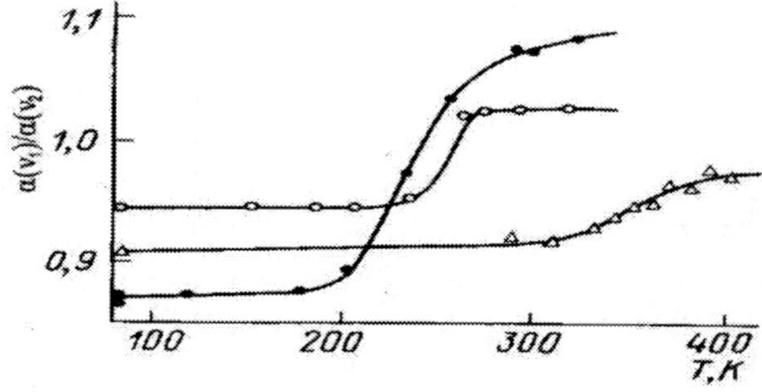

Figure 39. Temperature dependence of the absorbance ratio $\alpha\ (\nu_1)/\alpha\ (\nu_2)$ in doublet of skeletal band for PMMA (○), PAN(Δ) and PVC (●).

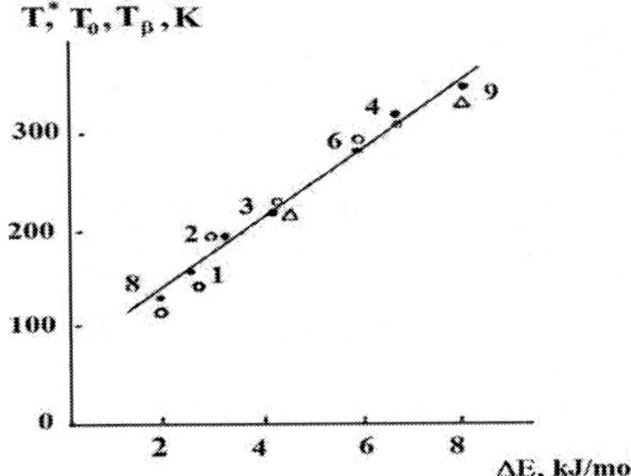

Figure 40. Correlation of the temperatures T^* (Δ), To (○) and T_β (●) with the energy difference $\Delta E = E_g - E_t$ for different linear polymers, Ref. [33].

It is noteworthy that the temperatures of the onset of translational movement of chain sections (in low-frequency Raman spectra) [66] and the temperature of a sharp increase in the vibration amplitudes of units in polymers, leading to a change in the conformation of macromolecules under the action of mechanical stress [158], were also close to the T^* values.

The close correspondence found in [33] seems to be especially important temperatures T^* and T_β, the characteristic temperatures in glassy polymers - T_0 -constant in the Fulcher – Vogel – Tamman (FFT) equation and temperature T_2 in the Gibbs – Di Marzio thermodynamic theory of glass transition [159, 160]. As expected for T_2 (159) and close to it meaning T_0, the temperature T^* meets the condition of termination of conformational rearrangements in chains and depends on the difference in energies of isomeric states ΔE (Figure 40).

The bands of torsional skeletal vibrations in the FIR spectra of PVC, PMMA, and PAN characterize vibrations in the chain around the equilibrium position in the potential well, which corresponds to a more stable (*trans*) conformer at low temperatures. It is characterized by a single band at ν_2 (Figure 38). With an increase in temperature, the amplitude of these oscillations increases and there is a successive excitation of torsional states, which ultimately leads to the overcoming of the internal barrier at T^* its rotation and population of a potential well corresponding to a higher-energy (*gauche*) conformer (see scheme). With a sufficient population of the potential well corresponding to the (*gauche*) conformer, additional absorption appears in the skeletal vibration band: the low-frequency component of the blow at ν_1.

Thus, T* is the quasi-thermodynamic temperature, starting from which correlated torsional vibrations of links in chain sections can reach amplitudes sufficient for the rotation-isomeric (*trans* ↔ *gauche* transition), i.e., for the manifestation of β-relaxation. The kinetics of β-relaxation is known to be determined by the height of the potential barrier for the rotational-isomeric transition and the proportionality of T_β to the value of ΔE (Figure 40) seems at first glance unexpected.

However, there is no sharp boundary between thermodynamics and kinetics of rotational-isomeric transitions in macromolecules. The thermodynamic characteristics that control the ability of macromolecules to change their conformation are related to the height of the total potential barrier and, therefore, to the kinetic properties of macromolecules. Following Volkenstein [112], the mean cosine of the angle of hindered rotation, which determines the magnitude of the statistical segment, as a measure of the flexibility of the chain, is equal to $Cos\ \varphi = [1 - exp\ (-\Delta E/kT)]/[1 + 2\ exp\ (-\Delta E/kT)]$. On the other hand, according to Bresler and Frenkel [161]: $Cos\ \varphi = Cot\ (E_0/kT) - 2kT/E_0$, where E_0 is a complete barrier for the rotational-isomeric transition. Comparison of these expressions shows the relationship between the size of the statistical segment, the height of the full potential barrier and ΔE.

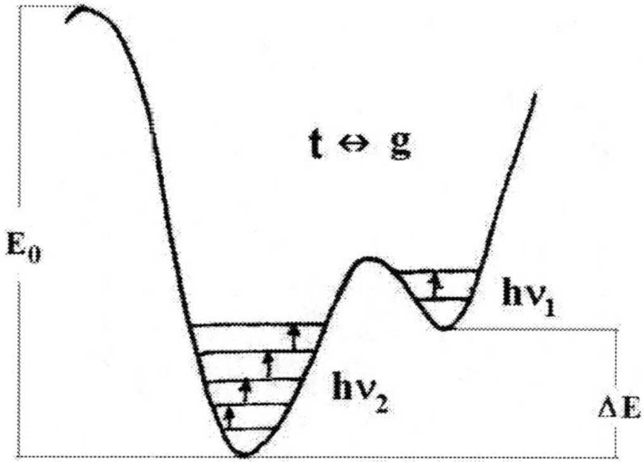

Figure 41. Scheme of sequential activation of torsional vibrations leading to a change in the conformation of a section of the polymer chain.

The thermodynamic rigidity of the chain, together with the energy barriers of inter- and intrachain interactions, also determines the β-transition temperature. Thus, for a large number of flexible-chain linear polymers, DSC showed the validity of the dependence $Q_{\beta} = (0.3 \pm 0.05)\ E_{coh}S + Q_0$ and, therefore, in accordance with the Arrhenius relation, $T_{\beta} = Q_{\beta}$ (kJ/mol)/(0.25 - 0.019 log $\bar{\upsilon}$) [129, 147].

Consequently, a close relationship between the equilibrium and relaxation properties of macromolecules is revealed during the rotational-isomeric transition.

The result obtained confirms the mechanism proposed for β - relaxation, including the displacement during the trans - gauche transition of adjacent monomer units due to the accumulation of torsional vibrations in the chain section commensurate with the length of the statistical segment, which in this case prioritizes the status of the kinetic unit of the β - transition.

4.7. Gamma-Relaxation

In flexible-chain polymers, relaxation losses are often observed in the temperature range intermediate with respect to the α-transition and low-temperature β-transition. This is a manifestation of γ - relaxation, the origin of which is still controversial. The molecular assignment of this transition has been unambiguously proven in two specific cases. This is, first, a chair-chair conformational transition in the cyclohexyl ring of polycyclohexyl methacrylate and other cyclohexyl-containing polymers, which manifests itself at 190 ± 10 K and 1Hz [162]. Second, this is the backbone-independent movement of long alkyl side groups in polyalkyl methacrylates and other comb-like polymers [144].

In more general cases, the γ-transition is attributed to vibrations of end or side groups [144, 163-165] or to torsional vibrations in short sections of chains, which are realized without overcoming barriers to internal rotation [166]. The first explanation is unlikely, since the contribution of end groups in high molecular weight polymers should be negligible. And the torsional vibrations of the side groups correspond, as a rule, to significantly lower potential barriers than the activation energy of the γ - transition.

In [167], the origin of γ-relaxation in polymers was studied using FIR spectroscopy and mechanical relaxation spectra. The initial hypothesis was the hypothesis that, in addition to the above-considered librations of an individual link and correlated torsional vibrations that determine the conformational mobility of the main chain, torsional vibrations are possible that do not lead to overcoming the barriers to internal rotation and cover chain fragments smaller than the statistical segment. Their relative role in the movement should obviously increase in those cases when the movement of the latter is difficult or even impossible to realize. For example, due to crosslinking of chains at a distance between crosslinks $N < S$ or in the case of oligomers with an average chain length $n < S$. It was assumed that such torsional vibrations in chains determine, in most cases, γ-relaxation in hykochain polymers.

To test this hypothesis, the DIR spectra of the oligomers MMA and α-methylstyrene, as well as dosed cross-linked PS and PMMA. In addition, for cross-linked PMMA (copolymers MMA-DMEG), the DIC spectra were compared with the spectra of mechanical losses measured in the temperature range 100 - 450K at Hz.

Let's go back to Figure 32 and 33, as well as to Figure 12 and 36 and analyze the absorption in the vicinity of 120-150 cm^{-1}, intermediate in frequency between the absorption due to libration and the absorption associated with skeletal torsional vibrations covering the correlation portion of the main chain (Cunowski segment). It can be seen that absorption bands at intermediate frequencies become clearly pronounced in the spectra of oligomers and cross-linked polymers. Thus, in the series of PMSs with the transition from polymer to oligomers and shortening of the molecules of the latter, an absorption band appears and grows in intensity at 120–150 cm^{-1} (Figure 12a). A priori, it could be explained, in addition to torsional vibrations of chain fragments, also by vibrations of the terminal groups of oligomers, however, this is evidenced by both the lack of correlation between the concentration of these groups and the band intensity, and the increase in absorption in the same region in the FIR spectrum of the PS as cross-linking develops (Figure 32).

Similar changes were observed in the intermediate frequency range during crosslinking of PMMA chains or for its oligomers (Figure 33). The FIR spectra of PMMA and 50-mer MMA were practically the same, but already the 9-mer was characterized by an increase in absorption, while the 7-mer and the dimer had a sharp increase in absorption at 140-160 cm^{-1}. A new absorption band at 150 ± 15 cm^{-1} also grows in crosslinked PMMA with an increase in the degree of crosslinking of its macromolecules, and it follows from the analysis of the spectra of crosslinked systems that this effect is not associated either with vibrations in cross bridges or with network defects.

Suppose that the initial hypothesis is correct and that the absorption at intermediate frequencies, both in the indicated cross-linked polymers and in oligomers, is indeed due to the torsion-vibrational motion in short chain fragments with $N < S$ units.

According to [73, 167], the potential barrier of such a motion can be determined from the ratio of the frequency of skeletal torsional vibrations ν_{sk} and the frequency of intermediate absorption νs: $Q_{loc} = Q_{sk} (\nu_{loc}/\nu_{sk})^2$, where $Q_{ks} \approx Q_\beta \approx Q_0 + Q_{lib}S$ includes the libration barriers S of the links of the correlation section and the barrier internal rotation Q_0. Formally, $Q_{loc} = Q_0 + Q_{libr}N$ and $N = (Q_{loc} - Q_0)/Q_{loc}$. The values of Q_{loc} obtained in this way for the studied oligomers and cross-linked polymers are given in Table 4. Their comparison with the activation energies of the γ - transition shows that in all cases $Q_{loc} \approx Q\gamma$ and at the same time two neighboring monomer units are involved in the act of torsion-vibrational motion. Consequently, when segmental motion is disturbed (suppressed) in the main chain, limited skeletal torsional vibrations come to the fore, which determine γ-relaxation.

This conclusion is also supported by the symbatic nature of the changes in the FIR spectra of oligomethacrylates and crosslinked PMMA samples and the spectra of internal friction of the latter. As the PMMA molecules cross-link, simultaneous increase in absorption at 120-150 cm^{-1} and its decrease at 225-230 cm^{-1} (Figure 33). The same effect is observed when the PMMA chains are shortened upon passing to oligomers with $N = 2$ and 7, in the spectra of which absorption appears as a separate absorption band at 120–150 cm^{-1} (Figure 35).

Table 4. Comparison of parameters of low-frequency skeletal vibrations and parameters of γ-relaxation in glassy polymers [135]

Polymers	Q_{libr}* (kJ mole^{-1})	ν^*_{sk} (cm^{-1})	ν_{loc} (cm^{-1})	Q_{loc} (kJ mole^{-1})	Q_γ (kJ mole^{-1})	T_γ, K (~1 Hz)	N
Cross-linked PMMA	14	225	140–160	28–42	43 [140]	170	2
Oligomers of PMMA	14	225–230	160–190	42–56	43 [140]	-	2–3
Oligomers of PMS	13	245	120–150	32 ± 7	35 [109]	140	2
Cross-linked PS	11	245	140	28 ± 5	33 [189]	132	2

* From Ref. [72, 120].

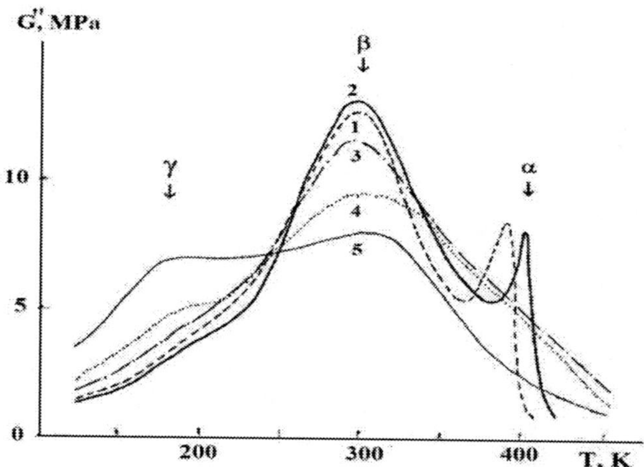

Figure 42. The loss modulus as a function of temperature for PMMA (1) and the MMA-DMEG copolymers with 5 mole % (2), 10 mote % (3), 50 mole % (4) and 100 mole % (5) DMEG.

From the spectra of mechanical losses of PMMA and MMA-DMEG (Figure 42) it can be seen that as the PMMA chains crosslink, the height of the β - relaxation peak ($T_\beta \approx$ 300K) also gradually decreases and the cooperative α - glass transition ($T \approx$ 400K), associated with intermolecularly cooperative movement of the same segments as in the β - transition. At the same time, the intensity increases symbatically γ - relaxation ($T \approx$ 170K).

The aforementioned increase in absorption in the intermediate frequency range in the FIR spectrum of a PS upon its chemical crosslinking is also accompanied by an increase in mechanical losses at T = 140–160 K, i.e., in the region of γ - relaxation [168]; the same was observed during radiation crosslinking of PS [169].

In Figure 43 summarizes, as applied to the PMMA system, the dependences of the intensities of the DIC absorption and relaxation characteristics on the average distance between crosslinks N in the polymer or on the average length of oligomer molecules n. It can be seen that the picture is qualitatively the same: it is in those cases when the values of N or n become smaller than the values of the statistical segment (for PMMA S = 6), a decrease in the intensities of β-relaxation and absorption is observed at 225-230 cm^{-1}. And, at the same time, the intensities of γ-relaxation and absorption increase at 140-190 cm^{-1}.

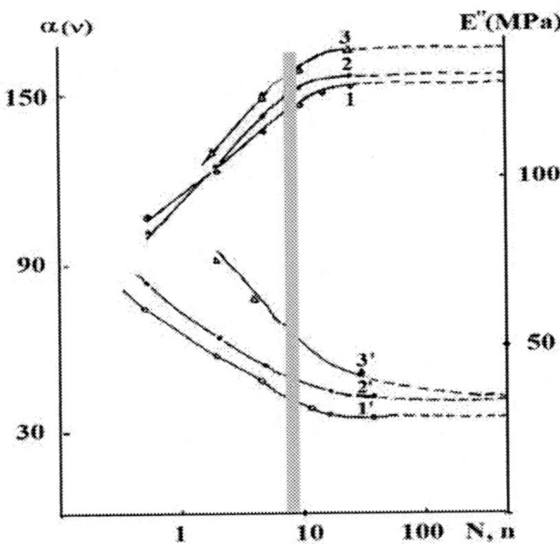

Figure 43. Intensity α(ν) of the skeletal bands at 225 cm-1 (1), 140 ± 20 cm^{-1} (1') in far infrared spectra of MMA-DMEG copolymers and at 225 cm^{-1} (3), 175 ± 10 cm^{-1} (3') in far infrared spectra of oligomethacrylates and the loss modulus G"max (in the ranges of *β*- (2) and *γ*- (2') -transitions) as a function of mean spacing between chemical cross-links expressed in the number of monomer units (*N*) or average length of oligomer molecule (*n*), Refs. [114, 135].

Finally, the estimate of the activation volume of this transition also testifies in favor of the considered concept of the *γ*-relaxation mechanism. In the case of PMMA, for the *β*-transition $V_a \approx 0.9$ nm^3, which approximately corresponds to the volume of the statistical segment of this polymer [170], while for the γ-transition $V_a \approx$ (0.4-0.5) nm^3, which approximately corresponds to 2–3 monomer units (171).

The role of crosslinks contributing to the manifestation of *γ*-relaxation, apparently, can also be played by rather strong quasi-chemical intermolecular sites, for example, created by hydrogen bonds. Thus, the γ-transition in polyamides and polyurethanes observed at 120–160K (1 Hz) is associated with the torsion-vibrational motion of short polymethylene chains between such sites [144], which does not contradict the stated model. Moreover, bridges of hydrogen bonds, obviously, can be created by sorbed water, which also leads to an increase in γ-relaxation, as in the case of polymethacrylates [172].

Thus, the data presented testify in favor of the initial hypothesis and the existence of a fairly general mechanism of γ-relaxation in flexible-chain polymers as local torsional vibrations in the main chain.

All the results presented in this chapter show that FIR spectroscopy makes it possible to specify the mechanisms of molecular movements responsible for relaxation δ, β and γ transitions in amorphous polymers, as well as to establish their relationship with the main molecular characteristics and structure. monomer unit, cohesion energy and thermodynamic rigidity of the macromolecule.

Chapter 5

Low-Frequency IR Spectra as a Source of Direct Information about Interchain Interactions in Polymers

5.1. Hydrogen Bond

Hydrogen bonding refers to the manifestation of the strongest secondary interactions in glassy polymers. The features and magnitude of these interactions are significantly influenced by the conformations and packing of macromolecules.

Although the energy of a hydrogen bond is weak compared to a covalent bond, this type of molecular interaction is sufficient to noticeably change the frequency and intensity of the vibrational spectrum of the polymer. In fact, these changes are so significant that IR spectroscopy becomes an important criterion for the presence of an H - bond.

Hydrogen bonding involves the interaction between a proton donor (group of the R_1 – X– H type) and a proton acceptor (group of the Y – R_2 type) and can be described schematically as R_1 –X– H ... Y - R_2. The formation of the H-bond is facilitated by the high electronegativity of the X and Y atoms, as well as their relatively small radius (for example, in the case of O, N, F) and the lone electron pair of the Y atom. Due to hydrogen bonding, the frequencies of the valence ν (XH) and ν (YR_2) oscillations decrease, while the deformation frequencies caused by the motion of H and Y atoms perpendicular to their X - H and Y - R_2 bonds, increases. The energy of hydrogen bonds is directly reflected in ν (X - H ... Y) valence and δ (X - H ... Y) deformation vibrations. The frequencies of these vibrations lie in the long-wave infrared (FIR) range.

Direct analysis of low-frequency vibrations of H– bonds in polymers is rare. On the other hand, considerable material has been accumulated on the FIR spectra of liquids with H– bonds (carboxylic acids, alcohols, phenols, etc.) [14, 173], due to which the task of assigning the H – bond vibration bands in the FIR spectra is significantly facilitated. For example, for low molecular weight systems, it was found that in aliphatic alcohols and phenols, stretching vibrations of H - bonds are manifested, as a rule, in the range 110–180^{-1}, and in carboxylic acids at 190–250 cm^{-1}. Bending and torsional vibrations of H - bonds are characterized by frequencies of 100 - 150 cm^{-1} and 40 - 60 cm^{-1}, respectively (174).

Let us turn to the FIR spectra of polymers with H– bonds. The FIR spectra of a series of polyamides (PA-2, -4, -6, -8, -10, -12), differing both in the number of CH2 - groups between amide groups and in the method of chain packing, are shown in Figure 44. They were studied in [175). Basically, polyamides exist in 2 crystalline modifications called α and γ. In the α-modification, the molecules form simple *trans*-chains, and in order to create the maximum number of H-bonds, the chains in this case must be antiparallel. In the case of the γ-configuration, the plane of CONH groups perpendicular to the CH2 groups and the chains in this form are parallel. For polyamides PA2, PA4 and PA6, the α-form is usually characteristic, while for PA8, PA10 and PA12 this is mainly the γ-form. Complete interpretation of the spectrum using group-theoretical analysis was performed only for PA6 (176). The assignments made in [175] for other polyamides are based on simpler models and are mainly qualitative in nature.

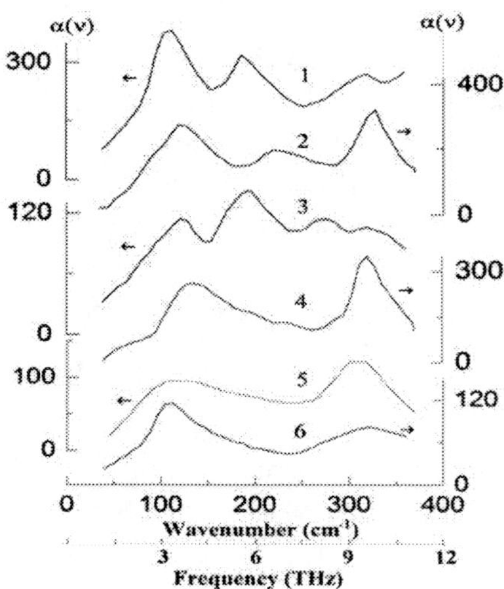

Figure 44. FIR spectra of polyamides: (1)-PA4, (2)-PA6, (3)-PA2, (4)-PA8, (5)-PAl0, (6)-PA12.

The frequency of the stretching vibration ν N - H of the amide group not bound by the H – bond is lower, than bound, since the hydrogen bond of the

H atom with the neighboring group leads to a decrease in the force constant (f_1) of the N - H vibration. This allows the spectra in the middle IR region, calculate f_1 = 704 dyn/cm, as well as the force constant vibrations of the amide group as a rigid fragment along the H-bond f_2 = 56 dyne/cm [178]. This translational vibration of the amide group in the direction perpendicular to the direction of the chain, as well as its torsional vibration with the force constant f_3 are shown in the scheme.

Assuming the translational vibration in the direction perpendicular to the main chain to be free, we have $\nu_{CO...H} = 130\ (f_2/2M)$, which at $M = m_O + m_C + m_N + m_H$ = 43 a. e. m. gives the calculated $\nu_{CO...H} = 105\ cm^{-1}$.

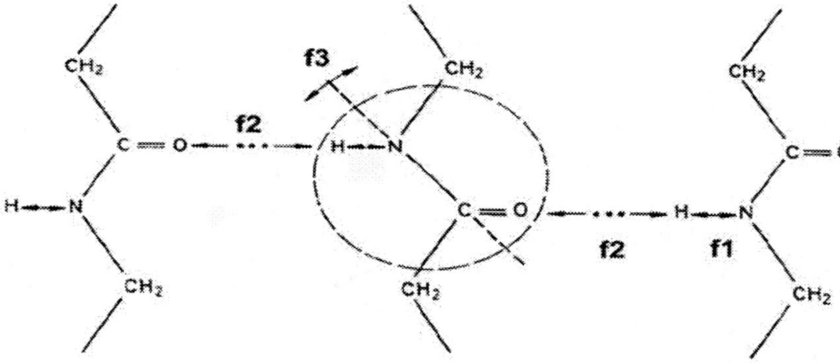

Thus, in the FIR spectra of polyamides (Figure 44), the band with a maximum at 100 - 110 cm^{-1} (in PA 2 polyglycine at 116 cm^{-1}) can be attributed to the translational vibration of the amide group along the H - bond. Deuteration of the amide group in PA6 did not change the position of the band, confirming this assignment [179]. A similar assignment was made for polyurethanes [180].

Higher-frequency absorption bands in the spectra of polyamides (in the spectra PA6 at 291 cm^{-1}, in PA 4 at 374 cm^{-1}, or at 630 cm^{-1} in PA 2) were attributed to torsional vibrations of the amide group on H-bonds. The librational origin of these bands is indicated by the correlations of their spectral parameters with the mass, moment of inertia, and dipole moment of the amide group.

Based on the above assignment, let us consider the changes observed in the spectra of polyamides (PA6 and PA12) after their preliminary deformation (Figure 45).

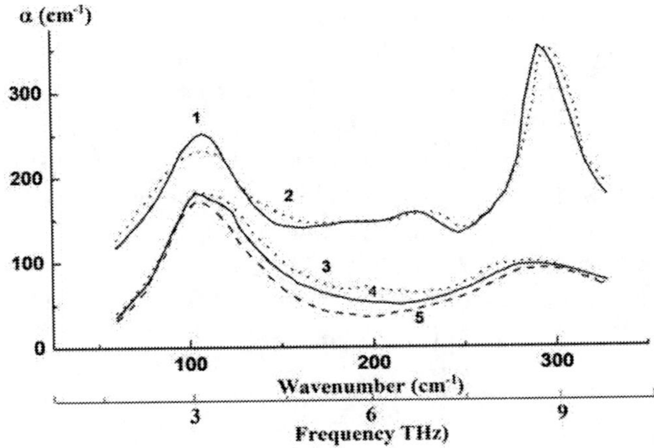

Figure 45. FIR spectra of PA6 (1, 2) and PAl2 (3-5) in initial state (t, 4) and after prestraining by 20-30% (2, 3); (5)-the spectrum of PAl2 under tensile stress $\sigma = 27$ MN/m^2.

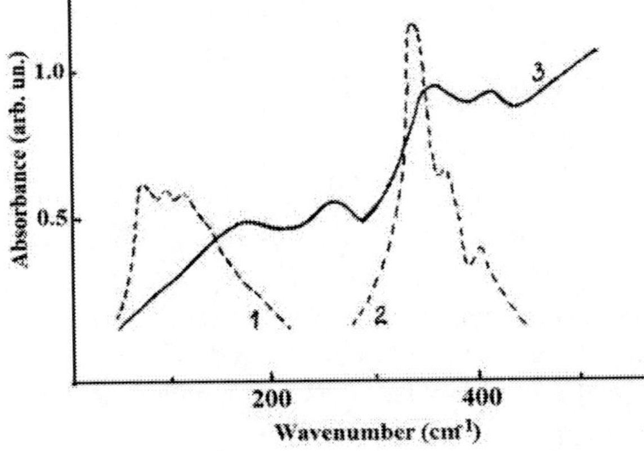

Figure 46. FIR spectra of MAA (Curves and 2; samples of thickness 200 and 90 micron, respectively) and PMAA (Curve 3; sample of thickness 200 micron).

The breaking of H - bonds in PA6 pre-deformed by 30% is evidenced by a decrease in the intensity and broadening of the absorption band in the range 100 - 110 cm^{-1}, which corresponds to the natural vibrations of H - bonds between NH and CO groups in this polymer. In addition, the half-width of the band at 295 cm^{-1} increases, reflecting the enhancement of the librational motion of macromolecules in the pre-deformed sample. An increase in the chain mobility upon deformation of PA12 is indicated by changes in the range of 150 - 280 cm^{-1}, which characterizes the torsional vibrations of polymethylene sequences including the CONH group. However, it is possible to catch the breaks of H - bonds in this polymer (band at 110 cm^{-1}) only by recording the absorption spectrum of PA12 under a tensile load. The FIR spectrum of PA12 under load also shows a narrowing of the bands due to translational and torsional vibrations associated with a decrease in the amplitude (anharmonicity) of these movements and an increase in the rigidity of the chains under load due to the so-called "mechanical glass transition".

In the FIR spectrum of polymethacrylic acid (PMAA) (Figure 46), containing intermolecular H-bonds, absorption bands were assigned based on its comparison with the spectrum of methacrylic acid (MAA) and other carboxylic acids [177]. In the MAA spectrum at $\nu \leq 400$ cm^{-1}, two groups of absorption bands are observed. The first of them - in the range of 300 - 400 cm^{-1} - can most likely be attributed to skeletal bending vibrations of the MAA molecule, similarly to what is considered for the spectrum of acrylic acid [181]. In the FIR spectrum of PMAA, these vibrations are apparently associated with absorption bands at 350 and 400 cm^{-1}. The second group of bands at frequencies in the range of 100 - 200 cm^{-1} should be attributed to the H-bond vibrations. It is known that in MAC H - bonds are capable of forming cyclic dimers, while three out of six vibrations of these dimers should be active in IR absorption: torsional, antisymmetric bending and antisymmetric stretching vibrations. The calculation makes it possible to associate the band at 136 cm^{-1} in the MAA spectrum with the stretching vibrations of the H – bond, and the band at 97 cm^{-1} with the deformation vibrations. Torsional vibrations can be the band is carried at 50 cm^{-1} [182]. If the absorption due to stretching, bending and torsional vibrations of H - bonds appear in the spectrum MAA in the range 100 - 150 cm^{-1}, then in the PMAA spectrum a similar absorption is shifted to a higher frequency region and is characterized by a wide structureless band with a maximum at 185 cm^{-1}. Note that the same shift and shape of the absorption band of H - bonds is also characteristic of a liquid, the molecules of which form polymer chains linked by hydrogen bonds [183]. It was found that when PMAA is heated to 500 K, the intensity of the band at 185 cm^{-1} decreases significantly, as does the ν_{O-H} bands, indicating that, under these conditions, most bridges from H - bonds in PMAA are destroyed [184]. The manifestation of H-bond vibrations at $\nu_{O...H}$ = 185 cm^{-1} in the FIR spectrum of polyvinyl alcohol (PVA) is also indicative. The absorption band at 360 cm^{-1} can be attributed to the manifestation of the overtone of this vibration $2\nu_{OH...O}$ in the FIR spectrum of PVA [5]. The identification of bands in the FIR spectra of PA, PMAA, and PVA was used to decipher the FIR spectra of a styrene-methacrylic acid copolymer [44, 185] (Figure 47).

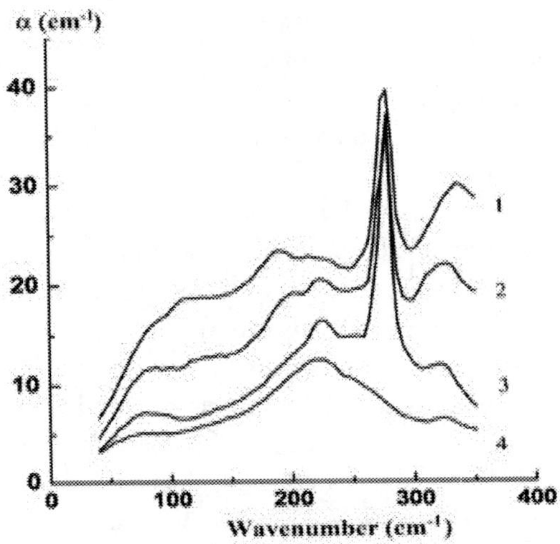

Figure 47. FIR spectra of styrene-methacrylic acid (MAA) copolymers. (1) -10% MAC; 2) - 5% MAC; (3) - 2% MAC; (4) - FIR spectrum of PS.

In these spectra, in addition to absorption related to polystyrene (PS) and absorption by skeletal deformation (band at 325 cm^{-1}) and torsional vibrations (band at 220 cm^{-1}), as well as by the libration of the benzene ring (band at ~ 80 cm^{-1}), one can also distinguish the absorption characterizing the H-bond in MAA units centered at 100 - 120 cm^{-1} and increasing with increasing MAA concentration in the copolymer. The absorption in the range of 150 - 200 cm^{-1} and 360 cm^{-1} indicates the formation of H - bonds of the polymer type in the copolymer. The relatively sharp peak at 273 cm^{-1} is apparently similar to the absorption at 290 cm^{-1} in the spectra of polyamides, i.e., represents torsional vibrations of COOH groups on H - bonds. In the PMAA spectra, the corresponding peak lies at 267 cm^{-1}.

Polyvinyl chloride (PVC) does not belong to polymers with H - bonds. However, the countable, based on normal coordinate analysis, assignment of absorption bands in the FIR spectrum of this polymer turned out to be effective only when taking into account intermolecular interactions of the C - H ... Cl - C type. As a result, the absorption bands at 64 cm^{-1} and 90 cm^{-1} were attributed to the translational and torsional vibrations of the C – Cl group on the H – bonds, and the band at 182 cm^{-1} to the torsion – deformation vibrations in the carbon chain [5]. At present, a more substantiated assignment of the bands at 64 and 90 cm^{-1} to the lattice modes has been proposed [71]; syndiotacticity of PVC.

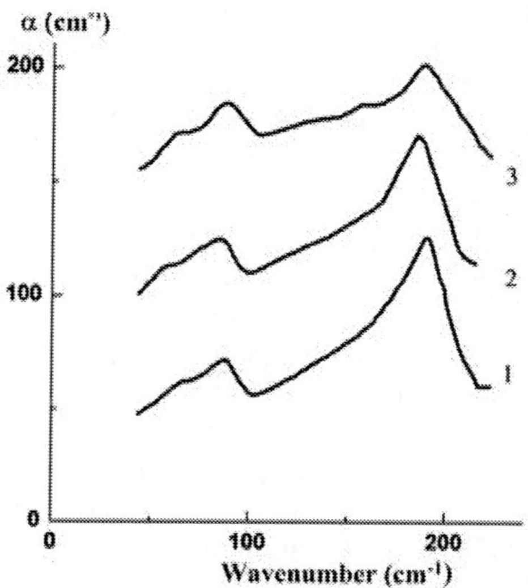

Figure 48. FIR spectra of plasticized PVC. (1) -PVC; (2) -PVC + 30 wt % DOP; (3) -PVC + 70 wt % DOP.

Confirmation of the decisive role of interchain interactions of the CH .... Cl - C type in the formation of the band at 182 cm^{-1} in the FIR spectrum of PVC was obtained in this work. Figure 48 shows the FIR spectrum of PVC plasticized with dioctyl phthalate (DOP). It can be seen that with an increase in the concentration of the plasticizer in the polymer, the main changes in the spectra relate to the intensity of the band at 182 cm^{-1}: it decreases with respect to the intensity of the bands at 64 and 90 cm^{-1}, while the latter only increase their half-width. Such changes in the spectra indicate a weakening or breaking of interchain bonds and an increase in the mobility of macromolecules in the plasticized polymer.

In general, the results of the above studies show the promise of using FIR spectroscopy, the most direct method for studying molecular interactions of the H-bond type in polymers.

5.2. Ionic Interactions

Ionomers - ionic polymers - have been the subject of the most intensive study, it has been established that unique mechanical and the rheological properties of these materials depend on the aggregation of ions included in the polymer chain [186, 187]. Ionomers are not only of scientific but also of significant industrial interest. On their basis, thermoplastic elastomers, polymer membranes, ion-exchange resins and polyelectrolytes have been created.

Most of the research works are devoted to ionomers, which are copolymers of ethylene, butadiene, and styrene with acrylic, methacrylic, or sulfonic acid at an ionic comonomer concentration not exceeding 10%. For such compositions, the low dielectric constant of the main component leads to the formation of ion pairs and further, through the formation of their multiplets, to the appearance of clusters [188].

The possibility of using FIR spectroscopy to study Coulomb interactions was shown back in the 70s. Then, intense absorption bands were identified in the region below 400 cm^{-1}, the connection of which with the vibrational motion of ions was easily traced due to the dependence of the position of the maximum of these bands on the mass of the cation [189–193].

The sensitivity of the vibration frequency of cation-anion pairs to the mutual arrangement of ions makes it possible to study the structure of ionic aggregates in polymer systems containing a small amount of salt groups in the chain, when the use of other methods is ineffective. The dependence of the value of the cation-anionic interaction on the degree of clustering, hydration, and the nature of the cation and the anionic part of the ionomer was shown in [194, 195] by the FIR spectra of ion-containing copolymers of styrene with salts of acrylic (PSAA), methacrylic (PSMK), and styrene-sulfonic acids (PSSK). Spectra of PSAA ionomers with chemically chemical composition, which can be represented as

$$-[CH_2-\underset{C_6H_5}{CH}]_n-[CH_2-\overset{CH_3}{\underset{COO^{(-)}M^{(+)},}{C}}]_m-$$

where M is the metal cation (Li, Na, K and Cs), and the value $m/m + n$ represents the molar fraction of carboxyl groups, are shown in Figure 49. The main feature of these spectra is the presence below 500 cm^{-1} of a broad, well-defined and intense band, which is absent in the spectrum of the nonionized copolymer. This band is centered at ~ 250 cm^{-1} for the Na + ionomer, at ~ 180 cm^{-1} for the K + ionomer, at ~ 115 cm^{-1} for the Cs + ionomer, and has a maximum at 460 ± 5 cm^{-1} for the Li + ionomer, not shown in the figure.

Since the position of these bands strongly depends on the mass of the cation, they can be attributed to the movement of the cation in the anionic field of the copolymer. The large half-width of the bands shows that there are several variants of the anionic environment of the cation, which differ in the strength of the Coulomb interactions. Figure 49 also presents the spectrum of the PSAA ionomer containing an increased concentration of sodium acrylate is shown. We see that with an increase in its concentration on the low-frequency wing of the main Na + band of motion, centered at 250 cm^{-1}, additional absorption appears at ~ 155 cm^{-1}.

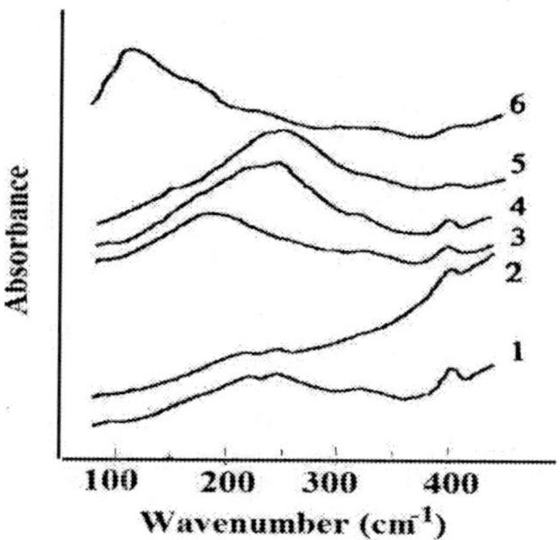

Figure 49. FIR spectra of styrene copolymers with 0.7 (1). 3.8 (4), 9.8 (5), mol.% sodium acrylate; (2), (3), (6) -, Li $^+$, K $^+$, Cs $^+$ ionomers.

The band at ~ 250 cm^{-1} - absorption caused by the movement of the cation - is present in the spectrum already at a low (<1%) concentration of Na + ions and can be attributed to the vibration of aggregates including several ions (low-order multiplet), while the band at ~ 155 cm^{-1} is apparently associated with vibrations of aggregates containing many cations -anionic cells. They can correspond to the formation of higher multiplets or clusters. Since in such ionic aggregates the interaction in the cation-anion cell is substantially screened by the environment, the vibration frequency of the cation is lowered in comparison with the vibration frequency in a simple multiplet.

Note that in the case of a styrene ionomer containing sodium methacrylate (PSMC), the absorption attributed to the formation of higher multiplets was found at ~ 170 cm^{-1}. Another difference in the PSMC spectra is the appearance of a new band at 380–390 cm^{-1}, which can be attributed to the bending vibration of the carboxyl group, which manifests itself in the presence of sodium ions. These discrepancies in the FIR spectra of PSAA and PSMK are apparently due to the difference in the structures of acrylic and methacrylic acids.

The assignment of the remaining bands with maxima at ~ 80, 218, 325, and 405 cm^{-1} or inflections at these frequencies in the spectra of all studied monomers PSAA and PSMK can be done by analogy with the assignment in the FIR spectrum of polystyrene. The lowest frequency band is the librational motion of benzene. rings, then 9B ring mode, carbon chain bending vibration and 16A ring mode.

FIR spectra of styrene copolymers with neutralized introduction of alkali and alkaline earth metal ions by polystyrene sulfonic acid salts (PSSK) are shown in Figure 50. In the range of 50-450 cm^{-1} of these spectra, the absorption bands associated with the movement of cations Na, K, Rb, Cs, Ca, Sr, Ba, are the most intense.

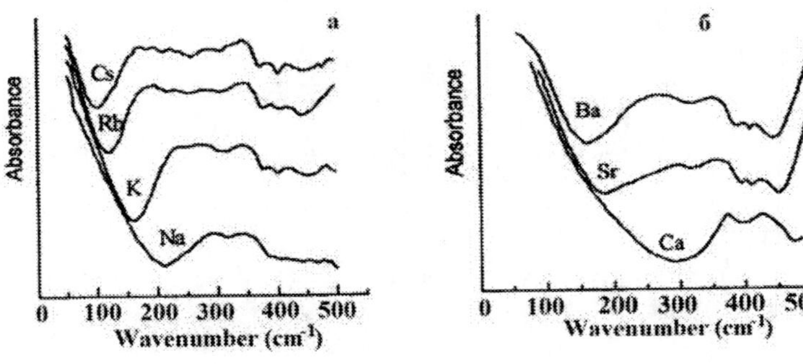

Figure 50. a) FIR spectra of Na^+, K^+, Rb^+, and Cs^+ of PSSA ionomers; b) FIR spectra of Ca^+, Sr^+, Ba^+ of PSSA ionomers.

The position of the maximum of these bands vmax is clearly determined by the mass of the M_c cation, and the dependences of ν_{max} on $(M_c)^{-1/2}$ for both alkaline and alkaline earth ions are practically linear and differ from each other only by their slope. Linearity means that the force constants of oscillations of ions belonging to the same group are close in magnitude. And they are larger for alkaline earth ions, since the slope of the dependence of ν_{max} on $(M_c)^{-1/2}$ is steeper here [196].

Of course, the identification of absorption bands associated with the vibration of the cation, based only on the fact that the frequency of these vibrations depends on the mass of the cation, is not entirely correct. First, because in the expression for the frequency of the oscillator ν (cm^{-1}) = $(1/2\pi c) \cdot (f/\mu)^{1/2}$, where μ is the reduced mass, must take into account the immediate environment of the cation. Second, the force constant $f = d^2U/dr^2$ at $r = r_{min}$, where U - the potential energy of interaction of ions, and r is the distance between them, will be determined by the nature of both the cation and the anion. The nearest anionic environment of the cation in the studied PSSA ionomers is oxygen atoms in tetrahedral or octahedral geometry. Assuming an octahedral environment, we have $\mu = (M_c \cdot m_0)/(2\ m_0 + M_c)$, where m_0 is the mass of the oxygen atom.

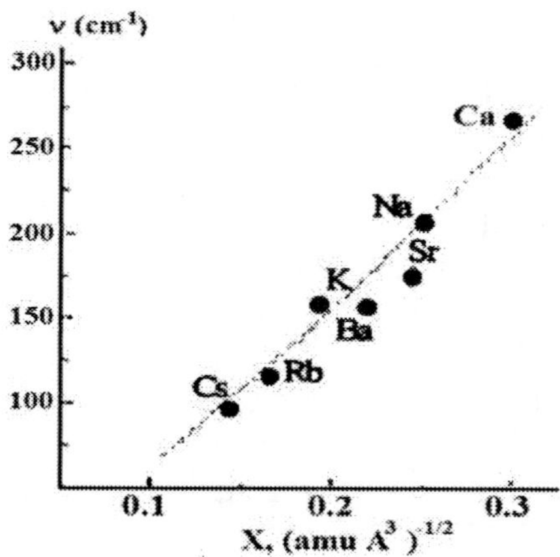

Figure 51. The dependence of the parameter X on the frequency of vibrations of the cation in the PSSA ionomer.

The potential energy of ionic interactions is described by the relation $U = -(Q_a \cdot Q_c \cdot e^2)/r + exp\,(-r/\rho)$, where Q_a and Q_c are the values of the charges of the anion and the cation in units of the electron charge (e), respectively, and ρ is the constant equal to 0, 33Å. The first term of this expression characterizes the Coulomb (electrostatic) attraction between opposite charges, the second - exchange repulsive forces.

The second derivative of the expression for the energy of interionic interactions (d^2U/dr^2) at $r = r_{min}$ gives $f = Q_a \cdot Q_c \cdot e^2\ (r - 2\rho)/\pi r^3$. With such f and μ, the vibration frequency of the cation is the position of the absorption band maximum - $\nu_{max} = [Q_a \cdot e^2/(2\pi c)^2]^{1/2} \cdot [Q_c(r - 2\rho)/\pi\mu r^3]^{1/2}$.

The obtained dependence for v_{max} can be represented as a linear equation $v = mX + b$, in which the parameter X will be determined only by the nature of the cation, and the slope of the dependence m - by the effective charge of its anionic environment.

The graph of v versus X is shown in Figure 51. It is seen that the frequencies of the maxima the vibration bands of the cation in the presented FIR spectra fit well a straight line, confirming the assignment made. The legitimacy of using the potential of ionic interactions makes it possible to have quantitative data on such (mainly Coulomb) interactions in ionomers.

How the Coulomb interactions change can be traced from the FIR spectra of ionomers, for example, during dehydration, which, as is known, strongly affects the structural, mechanical, and thermodynamic properties. In the FIR spectra of dehydrated ionomers, the vibration band of the cation shifts to low frequencies and noticeably narrows, which can be interpreted as a consequence of the enhancement of interactions between sulfonate groups and the cation upon removal of molecularly adsorbed water [197].

Table 5. Frequencies of vibrations of cations (cm^{-1}) in polysulfone ionomers

Cation	Average number of unsulfonated styrene units between sulfate groups			
	30	15	8	6
Na^+	220	210	–	210
K^+	170	162	170	155
Rb^+	120	120	120	116
Cs^+	100	100	100	95
Ca^{2+}	–	270	270	–
Sr^{2+}	180	177	175	170
Ba^{2+}	170	160	148	145

In the FIR spectrum of the ionomer annealed at $T > T_g$, the vibration band of the cation also narrows, but it is not possible to detect the shift of its maximum. This result can be explained by the compensating effect of the decrease in the strength of Coulomb interactions with increasing temperature and the impossibility of the formation of ionic clusters under these conditions. It is interesting in the light of the cluster mode to compare the spectra of ionomers with different concentrations of sulfonate groups. As shown in Table 5 the vibration frequency of the cation actually decreases with increasing concentration sulfonate groups, since the vibration of the cation will be determined by both an increased effective mass and a reduced force constant due to the screening of the cation-anionic attraction.

In the FIR spectra of composites of sulfonated polyetherurethane ($PEUS_i$) and K^+ ionomer of styrene-acrylic acid copolymer (ST/AA (K)) [198], the vibration band of the cation lies at 185 ± 3 cm^{-1} (Figure 52). Estimated the value ν_{max} = 180 cm^{-1}. With an increase in the ionomer concentration, the intensity and half-width of this band increase, and an additional absorption of 155 ± 3 cm^{-1} appears on its low-frequency wing, noting the appearance in the composites of higher-order multiplets - ionic clusters. The significant half-width of the band reflects the presence in the studied composites of cation-anion associates of different structures: from simple the ion pairs to their multiplets and clusters. Electron microscopy data confirm these conclusions. The presented results of studying the FIR spectra of ion-containing polymers indicate the broad possibilities of low-frequency spectroscopy for obtaining information on Coulomb interactions and the microstructure of ionomers.

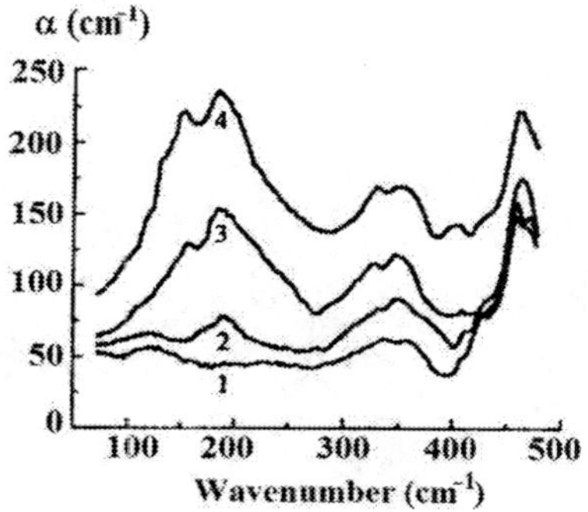

Figure 52. FIR spectra of composites PEUSi with K + ionomer ST/AA: - initial composite of PEUSi with ST/AA copolymer; 2 - K + ionomer 90 PEUSi: 10 ST/AA (K) wt.%; 3 - K^+ ionomer 70 PEUSi: 30 ST/AA (K) wt.%; 4 - K^+ ionomer 50 PEUSi: 50 ST/AA (K) wt%

5.3. Crystallinity, Hot Transitions, Spectra of Biological Macromolecules

Crystallinity bands arise due to the action of intermolecular forces in the crystal lattice of the polymer, where the molecules are packed in a regular manner. These crystalline bands are relatively rare in the IR spectra of mid-IR polymers, and such as the 730 cm^{-1} band in PE and 985 cm^{-1} in PS are just such bands [21].

The study of crystallinity, i.e., the volume fraction of crystallites in the polymer from IR spectra is also difficult because the IR spectrum is mainly determined by the structure and conformation of macromolecules; therefore, the assessment of the degree of crystallinity from the ratio of the intensities of the *trans* and *gauche* bands is not entirely correct, because disordered regions of the polymer can also have exactly the same the same shape.

FIR spectroscopy makes it possible to more directly investigate crystallinity of the polymer, since low-frequency vibrations will be intermolecular and are mainly determined by the packing of chains in the lattice.

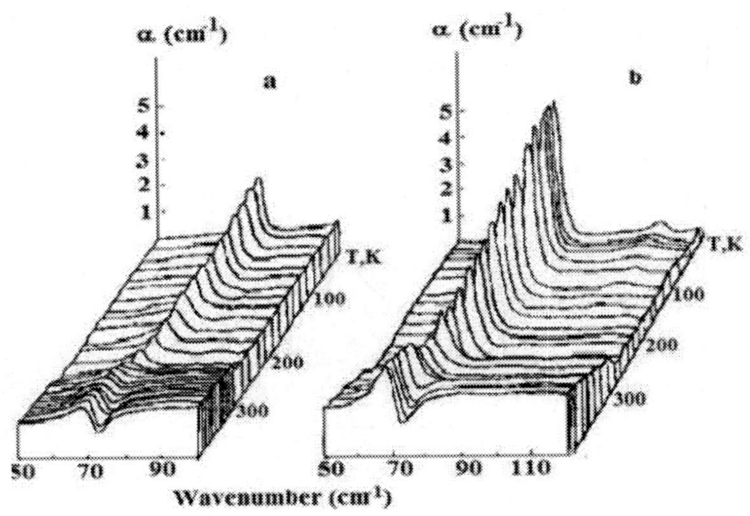

Figure 53. Change in far infrared spectra with temperature: (a) for low-density PE and (b) for linear high-density PE. From Frank and Leute, Ref. [4].

The effect of crystallinity (and temperature) will be illustrated by the following data. In Figure 53 a, b shows the absorption spectra and their temperature dependence for two types of PE. The result for low-density PE is suitable for prepared by introduction of branch points into the main circuit is shown in Figure 52 a. This polymer had about 40 such points per 100 carbon atoms, giving side chains of varying length. These branch points restrain crystallization, and in this type of PE there is a relatively low volume fraction of crystallites (~ 40%). Figure 53b shows the result for high density linear PE. The number of branch points in this case is only 3 per 1000 carbon atoms (hence the term linear chain), therefore, the degree of crystallinity is relatively high (about 80%). The difference in the degree of crystallinity of these samples, as we see, is reflected in different heights of the ν^a_5 mode peaks. The band at 108 cm^{-1} (mode ν^b_5) has a low intensity and appears only at low temperatures. With decreasing temperature, the band shifts to higher frequencies and can also be classified as a crystallinity band [199, 200].

For PPs of various degrees of crystallinity, a narrowing of the absorption bands in the FIR spectrum was found with an increase in the latter [20] (Figure 54). This effect is especially pronounced at low temperatures, in conditions of suppression of many phonon processes. An increase in intermolecular interactions in the crystalline regions of the polymer also leads to the splitting of some bands. With a decrease in the degree of crystallinity, the intensity of the crystallinity bands decreases and the intensity of the background absorption associated with disordered regions increases.

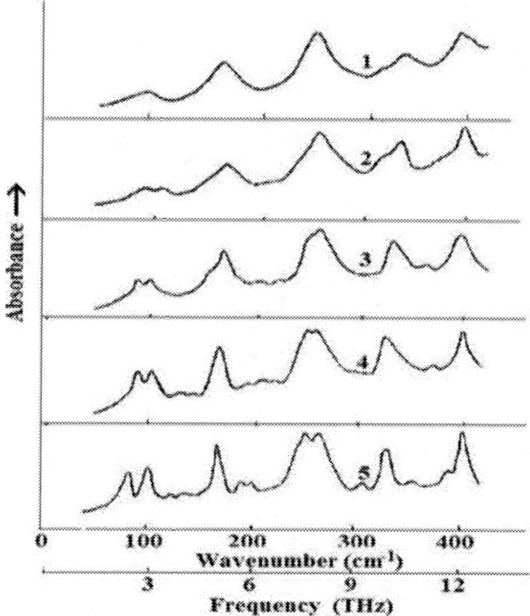

Figure 54. Far infrared spectra of isotaetic PP with different degrees oferystallinity: (1) 52%; (2) 58%; (3) 66%; (4) 70%; (5) 75%. From Goldstein et al., Ref. [20].

In the FIR spectrum of highly crystalline PTFE at low temperatures, there are four narrow absorption bands (Figure 3b, c), three of which at 46, 70, and 85 cm^{-1} correspond to translational lattice vibrations, and the band at 55 cm^{-1} appears as a result of the overlap of two closely spaced bands corresponding to lattice torsional vibrations [62].

It was shown in [201, 202] that FIR spectroscopy, when used correctly, makes it possible to trace microstructural transformations, for example, in polyvinylidene fluoride (PVDF) placed in a static field (1MV/cm) at 65^{0}C. Under these conditions, the transformation of the α-form of the polymer into the β-phase is observed at a rather low value of the applied field, and this can be traced by an increase in the intensities of the crystallinity bands at 70 cm^{-1} (lattice libration mode for the β-phase) and at 176 cm^{-1} (torsional vibration of the CH-CF-CH-CF group for the α-phase (Figure 55). The appearance of absorption bands corresponding to the crystalline state of the polymer in the long-wavelength IR region is also confirmed by the results of [203], in which the absorption spectra of polypropylene and triacetylcellulose in the wamorphous and crystalline state were obtained in the range 400–100 cm^{-1} (Figure 56). The spectra were measured using the first domestic long-wavelength spectrometer LIRS-1 [204].

Studies of crystallinity using FIR spectra are also discussed in papers [22, 205–207].

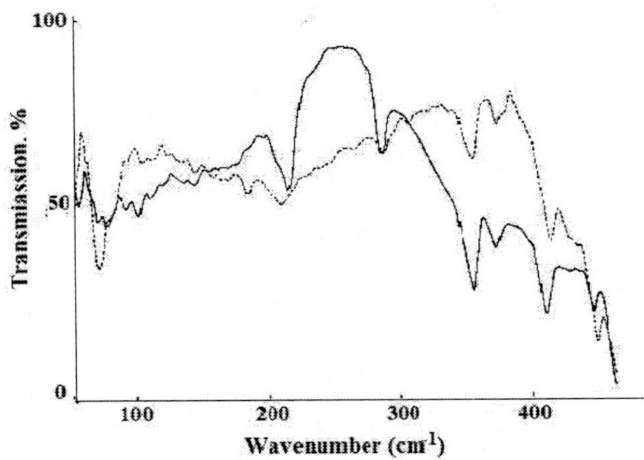

Figure 55. DIR spectra of PVDF in polarized light. Solid line: before polarization, dotted line after polarization.

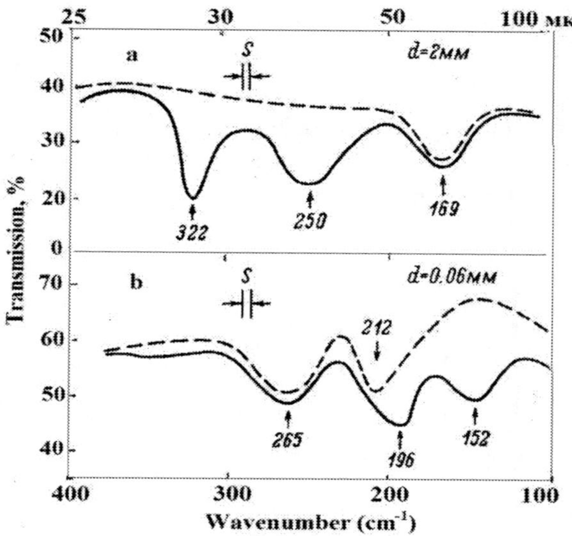

Figure 56. Absorption spectra of crystalline (solid curves) and amorphous (discontinuous curves) propylene (a) and triacetylcellulose (b).

In this review, we hardly touch upon the works on the assessment of the conformational structure of macromolecules using FIR spectra [208-210], since they do not contain any specificity in comparison with similar works performed using spectra in the mid-IR region. In the FIR spectra of polymers, a change in the conformational composition is reflected in the redistribution of the intensities of the components of the duet contour of conformationally sensitive absorption bands with a change in the polymer temperature, as well as as a result of orientational stretching, pressure, etc.

To be correct, it is important to emphasize another feature of FIR spectra. We are referring to the so-called "hot transition" bands, or simply hot bands. This absorption is made up of transitions between higher vibrational levels. Hot bands should be distinguished from overtone bands corresponding to transitions from the ground level to a higher level, which are very useful in studying the anharmonicity of vibrations [211]. If the use of overtones is associated with the expansion of the traditional IR range towards high frequencies, then the detection of hot bands requires moving into the FIR region of the spectrum, since here transitions between higher vibrational levels are much more likely.

Indeed, let n_1 and n_2 be the number of molecules at the main (l_0) and first vibrational levels (l_1), and ν is the frequency of the fundamental vibration and (T) is the temperature. Then $n_1/n_2 = exp\ (-h\nu c/kT) = exp\ (-1.43\ \nu/T)$. The estimate for $\nu = 150$ cm^{-1} at $T = 300$K gives $n_1/n_2 = 0.4$, i.e., the population of the first vibrational level is already high enough at room temperature that we can expect the appearance of bands due to "hot transitions" $l_1 \rightarrow l_2$. These bands should have slightly lower frequencies than the main bands due to transitions $l_0 \rightarrow l_1$ due to anharmonicity of low-frequency oscillations.

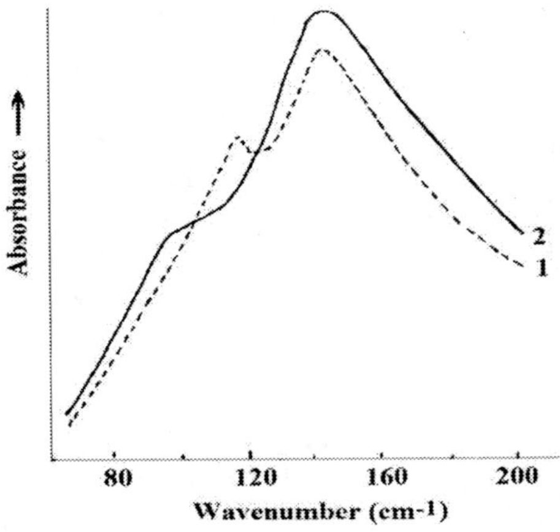

Figure 57. Band of skeletal torsional vibrations in the FIR spectra of 0.5 mm films of polydioxolane (1) and polytetramethylene oxide at room temperature [212].

In Figure 57 shows the FIR spectra of polydiooxalane (CH_2CH_2 OCH_2O)n and polytetramethylene oxide (CH_2 CH_2 CH_2 CH_2 O)n. An intense band with a maximum at ~ 145 cm^{-1} refers to torsional vibrations in the main chain of these polymers containing oxygen "hinges" [212]. For both polymers, the band at ~ 145 cm^{-1} contains satellites of lower intensity on the low-frequency wing, assigned to transitions of the $l_1 \rightarrow l_2$ type. The frequency difference between the main and hot bands is 20 cm^{-1} in the case of polydiooxalane, and 40 cm^{-1} for polytetramethylene oxide, i.e., the anharmonicity of torsional vibrations in the second polymer is noticeably higher than in the first. This is consistent with measurements of potential barriers to molecular motion by nuclear magnetic resonance for these polymers [213]. For unambiguous assignment of the considered low-frequency satellites of the band of torsional vibrations to the bands of hot transitions, however, temperature measurements are necessary.

Note that the significant population of the higher levels of low-frequency vibrations, expected already at room temperature, leads to the need to take into account the correction for stimulated emission in the FIR spectra. According to [214, 215], taking into account such an amendment should lead to significant changes in the FIR contours -the band at ~ 145 cm^{-1} contains satellites of lower intensity on the low-frequency wing, assigned to transitions of the $l_1 \rightarrow l_2$ type. The frequency difference between the main and hot bands is 20 cm^{-1} in the case of polydiooxalane, and 40 cm^{-1} for polytetramethylene oxide, i.e., the anharmonicity of torsional vibrations in the second polymer is noticeably higher than in the first. This is consistent with measurements of potential barriers to molecular motion by nuclear magnetic resonance for these polymers [213]. For unambiguous assignment of the considered low-frequency satellites of the band of torsional vibrations to the bands of hot transitions, however, temperature measurements are necessary.

Note that the significant population of the higher levels of low-frequency vibrations, expected already at room temperature, leads to the need to take into account the correction for stimulated emission in the FIR spectra. According to [214, 215], taking into account such an amendment should lead to significant changes in the FIR contours absorption bands and their shift to zero frequency. Following the opposite point of view [216] that this correction for stimulated emission is currently impossible, and taking into account some indirect evidence that its role is insignificant, in this review we proceed from the spectra given in the papers under discussion without introducing into them any amendments.

In conclusion, we present the FIR spectra of the most complex and important polymer objects: biological macromolecules. Most of the spectroscopic works concerning biomacromolecules deal with the wavelength region above 300 cm^{-1}, since such spectra are easier to interpret and due to the increase in experimental difficulties with decreasing frequency. Nevertheless, the low-frequency range is important mainly for the conformational study of biomacromolecules, since the vibrational modes that occur here are more sensitive to the environment than localized high-frequency vibrations.

In [217], the FIR spectra of amino acids and simplest polypeptides were studied. In Figure 58b shows the spectra of poly-L-alanine (PLA). In this spectrum, the absorption bands were assigned (at 248 cm^{-1}) to skeletal bending vibrations, at 205 cm^{-1} to torsional skeletal vibrations, and the 103 cm^{-1} band to the vibrations of the H-bond between CO and NH groups. The lowest-frequency absorption did not receive an unambiguous assignment, given the possible contribution of such factors as the presence in PLA of different (α-helix and β-form) conformations of macromolecules, the presence of flexible and rigid regions, heavy metal atoms, etc. In the FIR spectrum of poly-α-glycine (Figure 58a) a band at 203 cm^{-1} are associated with stretching of the H-bond between the C = O and OH groups. In this case, the band 117 cm^{-1}, intensity of which decreases with decreasing temperature, possibly refers to a hot transition. In [219], the absorption band at 120 cm^{-1} with a shoulder at 85 cm^{-1} in the spectrum of poly-L-alanine was attributed to bending skeletal vibration, and its arm to torsional vibration. The band at 45.5 cm^{-1} in polyglycine was assigned to the interchain mode.

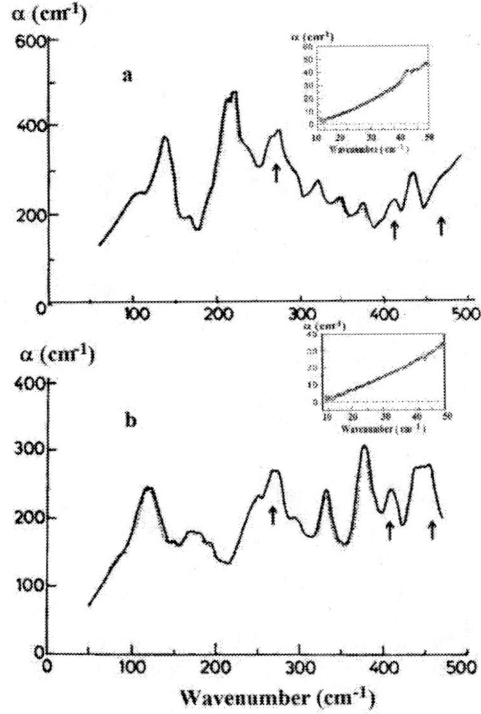

Figure 58. FIR spectra of films of poly-α-glycine (a) and poly-L-alanine at room temperature [217]. Arrows indicate solvent stripes. The insets show the spectra in the range 10-50 cm^{-1} [218].

Regarding other features of the FIR spectra of polypeptides, we would like to draw your attention to a sharp rise in absorption starting from

50 cm^{-1}, which is not quadratic as is expected for purely one-phonon disorder-induced absorption caused by acoustic modes. It indicates the existence of a large number of individual, more localized modes that cause inhomogeneous broadening of absorption in addition to background absorption.

In [220], the dispersion curves calculated for the right-sided α-helix of PLA are presented. The difficulty of using these data for assignment at low frequencies is due to the fact that, in this case, acoustic and optical which vibrations are not completely separated from each other.

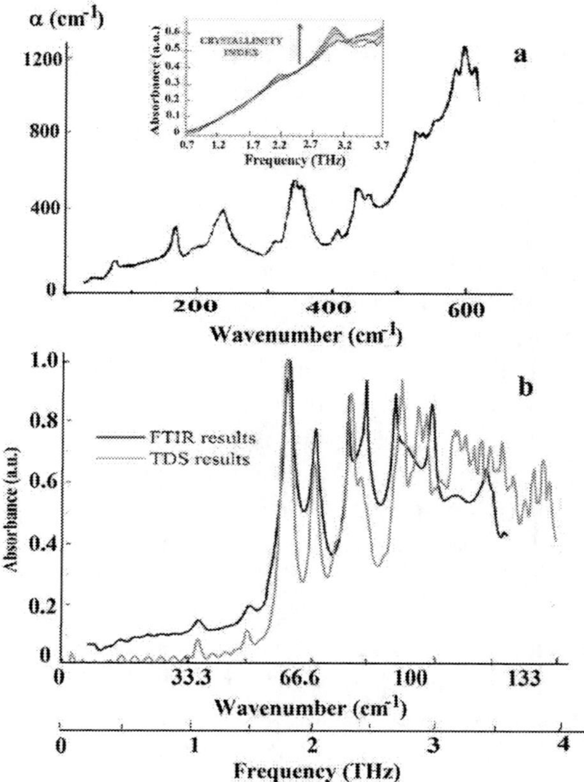

Figure 59. FIR spectra of cellulose (a) at 110K [223] and ascorbic acid (b) at room temperature [226]. The inset shows celluloses with different degrees of crystallinity in the range 0.2 - 4 THz [227].

FIR spectra of biopolymers are also discussed in [221-225]. As an example, we present the FIR spectra of cellulose polysaccharide and ascorbic acid monosaccharide (Figure 59b). Note that some of the bands in the spectrum of the polysaccharide can be interpreted in terms of the characteristic frequencies of the monosaccharide, which in turn can be calculated based on the normal vibrational analysis. Other spectra of polysaccharides such as starch, pectin, agar have a common property that their FIR spectra have a broad absorption band in the region of 200 cm^{-1}, associated with the presence of liquid H_2O in these biopolymers. This may also be associated with a sharp increase in absorption with increasing wavelength [223].

5.4. Far Infrared and Terahertz IR Spectra of Liquid Crystal Polymers

Low-frequency molecular spectroscopy is becoming more and more widespread in the study of the properties and structure of liquid crystals (LC) and liquid crystal polymers (LCP) [228-231]. So, along with traditional in recent years, research has been developing new spectral methods, which include terahertz molecular spectroscopy (see reviews [232, 233]). The interest of researchers working in these areas of spectroscopy in LC systems is due to the fact that one of the important mechanisms underlying the formation of low-frequency spectra has a pronounced intermolecular character. At the same time, the very fact of the existence of a liquid-crystalline state is due precisely to the forces of intermopecular interaction [234]. In addition, the low-frequency spectra of liquid crystals, on the one hand, and disordered systems, such as liquids and glassy substances, on the other, differ significantly. Since LC systems occupy, in a certain sense, an intermediate position in the series of these systems, it can be expected that low-frequency spectra will clearly reflect the structural features of LC systems. In particular, the study of such a characteristic property as dichroism in polarized low-frequency absorption spectra of oriented LC systems makes it possible to extend the concepts developed to explain dielectric loss spectra to the optical frequency region (see below). Consequently, it becomes possible from a unified point of view to describe the properties of liquid crystal systems in a fairly wide spectral range, which is in agreement with theoretical concepts [235].

Finally, studies of the low-frequency spectra of LCD systems are of great importance for a deeper understanding of the nature of these spectra, which is still not fully elucidated. Therefore, the problems that one has to face when studying LCPs using low-frequency spectra, both methodically and theoretically, are inextricably linked with the general problems of low-frequency vibrational spectroscopy.

Works on the study of the dynamics and structure of the LCP on this topic can be conditionally divided into two directions. One of them is devoted to the study by the methods of low-frequency IR spectroscopy of various phase states of LCPs in a wide temperature range (especially near the points of phase transitions) in order to obtain information on the structural features of the LC state within a given phase and during the transition from one phase to another, as well as dynamics of the process of phase transformations (see, for example, [236]). Another area includes works devoted to the study of the nature of absorption of the nematic and isotropic phases of LCP in the low-frequency region of the spectrum. These studies continue the cycle of works related to the study of the absorption mechanism of liquid systems in the long-wavelength IR region of the spectrum [74, 237].

Let us dwell on the analysis of the work devoted mainly to the study of the features of the local dynamics and the orientational state of rigid-chain liquid-crystalline polymers of the Vectra A950 and Armos type. Currently, the most close attention is paid to the creation and study of high-strength, thermo- and fire-resistant synthetic fibers for the modern high-tech industry. Domestic and foreign practice of their use has shown that fibers from lyotropic systems based on heterocyclic polyimides and aromatic paraaramides of the Armos type, as well as thermotropic systems based on liquid crystal (LC) copolyesters of the Vectra type, have the most record characteristics.

Despite the difference in the methods of obtaining these rigid-chain LC polymers, the high molecular orientation required for the realization of high values of strength and elastic modulus in both cases is achieved during the formation of fibers during spinneret drawing [238]. The main thing is that for a further (and significant) increase in strength, these fibers are subjected to the so-called heat treatment (HT) - heating in certain temperature and time modes. During HT, due to spontaneous self-organization, the supramolecular structure of the polymer improves, and degradation products are removed from the fibers, additional polycondensation, crystallization, and the formation of crosslinks [239].

For such large-scale rearrangements, highly oriented macromolecules must have mobility at least similar to segmental mobility in flexible-chain polymers. The elucidation of the molecular mechanisms of mobility in such systems is the subject of this work.

Earlier, in a series of works on the study of molecular mobility in LC polymers (LCP) by the method of nuclear magnetic resonance (NMR) of broad lines [240-243], it was found that large-scale rearrangements in the processes of self-organization and hardening of LC polymer fibers are provided by the specific cooperative movement of macromolecules in LC polymers. a condition called by the authors "quasi-segmental". This work is experimental, using IR spectroscopy in the middle and terahertz ranges in order to obtain additional information about the "quasi-segmental" mobility in such rigid-chain polymers.

Engaging terahertz IR spectroscopy (FIR) represents of special interest for the physical chemistry of LCP. IR-active degrees of freedom directly involved in molecular dynamics in this range correspond to high-frequency relaxation and high-amplitude torsional vibrations of chain fragments, as well as their correlated motion, which provides conformational rearrangement [136]. In addition, as already noted, it seemed that one of the main mechanisms underlying the formation of the FIR absorption spectra has a pronounced intermolecular character.

The objects of study were: 1) LC copolymer of the industrial grade Vectra A 950, having the formula

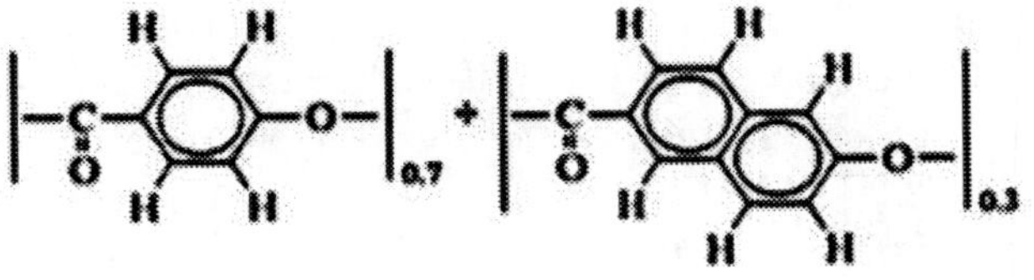

and containing 70 mol. % of 4-hydroxy-benzoic acid fragments and 30 mol. % 6-hydroxy-2-naphthoic acid and 2) Armos LC copolymer having the formula

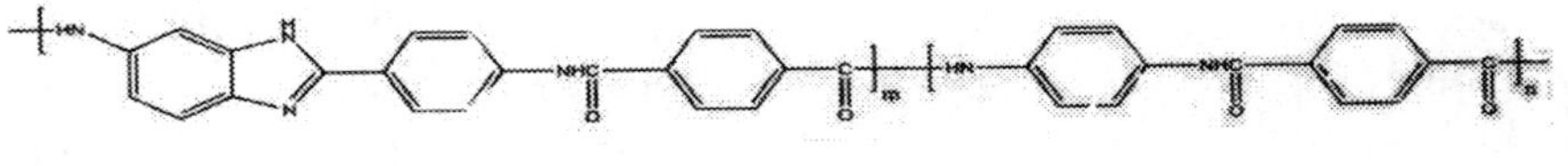

and formed by the polycondensation reaction of p-phenylene diamine and 5-amino-2- (p-aminophenyl) benzimidazole with terephthalic acid.

The main feature of the Vectra A 950 LC polymer is that it has two mesogenic groups in the main chain, which are not separated by flexible junctions. This imparts rigidity to the copolymer macromolecules, greatly reducing the molecular mobility of its chains [244]. Mesogens of the Armos copolymer macromolecules are also not separated by flexible junctions. Additional molecular stiffness providing high thermal stability and outstanding mechanical the properties of this copolymer are imparted by conjugated benzimidosole and phenolic rings in the main chain; and H-bonds between chains.

The study of the Vectra A950 copolymer was carried out on samples in the form of granules and fibers (filaments). Copolymer granules were obtained by molding a starting material from Hoechst-Celanese corp. Its fibers are made in a laboratory facility Physicotechnical Institute according to the methodology [245]. Samples of the Armos copolymer were obtained in the form of fibers: initial and passed HT.

The measurement of the IR transmission spectra of the investigated Vectra A950 and Armos samples was carried out on two instruments: in the mid IR region of 4000-400 cm^{-1} on a Perkin-Elmer 577 spectrophotometer with a resolution of 2-4 cm^{-1}, in the far IR region from 400 to 20 cm^{-1} on a Hitachi FIS 21 spectrometer with a resolution of 1-2 cm^{-1}. Spectra of granules were obtained on samples prepared in the form of tablets from a mixture of 1: 100 granules with PE. To obtain the spectra of the fibers, the filaments of the sample were regularly placed on a 20x12 mm frame. The polarizers during operation in the mid-IR were gratings on a substrate made of fluoroplastic (4000-1150 cm^{-1}) and polyethylene (1150-400 cm^{-1}), and in the far-infrared (FIR) polarization of radiation was carried out using a polyethylene replica of a diffraction grating with 1200 pcs./mm.

Another research method was proton magnetic resonance (PMR). It made it possible to track both the large-scale motion of macromolecules at high temperatures and changes in the structure of LC polymers. The PMR spectra were recorded in the temperature range from 20 to 300^{0}C on a broad-line NMR spectrometer developed and manufactured at the Ioffe Institute[246].

The IR spectra of Vectra A 950 granules and filaments in the range 4000-400 cm^{-1} are shown in Figure 60. The assignment of bands in these spectra was done in [244, 247]. Spectra of the mid-IR range (including polarized spectra) of these systems turned out to be practically the same, apparently because the short-range intramolecular forces in the filaments do not differ significantly from those in the granules.

The recorded long-wavelength FIR spectra of Vectra A 950 granules and filaments in the range of 400-20 cm^{-1} are shown in Figure 61.

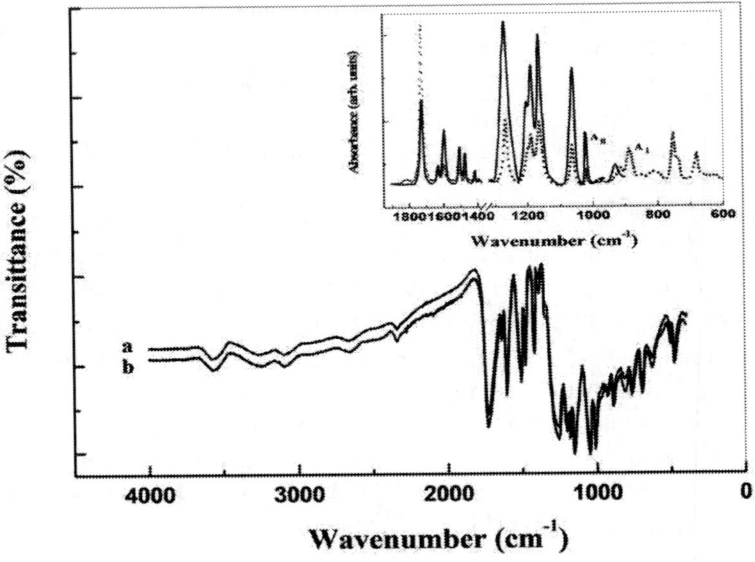

Figure 60. IR spectrum of Vectra A 950. Granules-a, filaments-b. For clarity, the spectrum of filaments is shifted along the ordinate. Inset: polarization absorption spectra of Vectra A 950.

Literature data on identification of bands in the FIR spectrum of Vectra A950, apparently not. There are only data concerning low-frequency absorption in low-molecular-weight LCs, but they are quite scattered and refer mainly to low-molecular nematics of the MBBA type (4-methoxybenzylidene-4-butylaniline) with benzene rings as mesogens and alkyl groups as solutions [248]. Information on the polarization FIR absorption spectra of some or there are practically no LC polymers.

To identify absorption bands in the FIR spectrum of Vectra A 950, we used the experimental and calculated literature data on the long-wavelength spectra of hydroxybenzoic acid [249] and hydroxy naphthoic acid [250], which are part of this copolymer. Based on these values, the preliminary assignment of the bands, as well as their dichroic ratio, obtained from our polarization measurements on Vectra A 950 filaments and pellets is presented in Table 6.

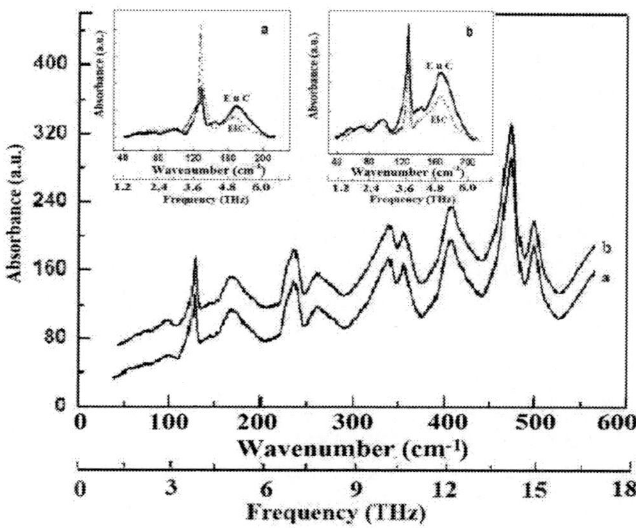

Figure 61. FIR spectra of Vectra A 950. The spectrum of filaments is shifted along the ordinate. Inset: polarized absorption spectra of Vectra A 950 granules (a), filaments (b) in the range 40-220 cm^{-1}.

Table 6. Spectral characteristics of the FIR absorption spectra of the Vectra A 950 LCP

Vibrational modes Vectra A950, (cm^{-1})	Approximate description	Dichroic ratio	
		$R_{granuls}$	R_{fibers}
69	*N*- ring libration	0.75	0.95
96	*B*- ring libration	0.95	1.15
125	Poley absorption + COO torsion	0.45	1.55
143	ν_{10b} *B*- ring	1.65	1.55
169	'Batterfly' *N*- ring	1.50	1.55
235	CCC def.	1.42	1.33
262	COC torsion	1.55	1.43
338	N_{10B} *N*- ring	1.55	1.43
355	'Envelope flap' *N*-ring	1.65	1.55
407	ν_{16a} *B*-ring	1.20	1.25
475	*N*-ring def. (B_{3u})	1.25	1.33

Symbols: F-fibers; G-granules; N-naphthalene; B-benzene; def.- deformation.

To estimate the orientational order parameter of the copolymer $S = (3 <Cos^2\Theta> -1)/2$, where Θ is the angle between the axis of the molecule and the axis of the fiber, the *formula* $S = (R-1)\ (R_0 + 2)/(R + 2)$ was used (R_0-1), where R is the dichroic ratio, $R_0 = 2Cot^2\psi$, and ψ is the angle between the moment of transition and the molecular axis of the chain, parallel in this case to the direction of drawing. The angle ψ for the main components of Vectra A 950 - benzene rings and C-O-C- bonds - is not exactly known, but in the first approximation it can be taken equal to 0^0, and then $S = (R-1)/(R + 2)$. The dichroic ratio is determined was calculated as the ratio of the absorption of radiation polarized parallel to the molecular axis of the LCP (and the direction of drawing) - $A_{//}$ to the absorption of radiation polarized perpendicular to the direction of drawing -$A_{\perp}$, i.e., $R = A_{//}/A_{\perp}$. In turn, $A_{//} = ln\ (1/T_{//})$, and $A_{\perp} = ln\ (1/T_{\perp})$, where $T_{//}$ and $T_{\perp}$ are the transmission of polarized IR radiation with an electric vector (*E*) parallel and perpendicular to the drawing direction (*C*) (and slots of the device).

A comparison of the dichroic ratios of absorption bands in the FIR spectra of granules (R_g) and in the spectra of fibers (filaments) (R_f) shows that the values of R_g and R_f do not differ much from each other, with the exception of the three lowest frequency bands characterizing the strong vibrations (librations) of mesogens (benzene and naphthalene rings of the copolymer). The R_g and R_f values (and the orientational order parameters) obtained for the absorption band at 127 cm^{-1}, which is considered here as absorption by the Poley mechanism, are especially different. Recall that the term "poly absorption" is used within the framework of the concept most widespread and currently used for the interpretation of low-frequency IR spectra of condensed media [136], which is based on the notion about librating molecules (here –mesogens), which are in the field of intermolecular forces. In this case, in addition, it is taken into account that the formation of the low-frequency spectrum can involve polar fragments of the LC polymer capable of performing torsional motions, as well as other possible intramolecular reorientations. We are talking about the contribution to the intensity and profile of the absorption band at 127 cm^{-} vibrations of the COO group, which binds mesogens and contributes to the conformational mobility of polymer chains.

Such intramolecular vibrations are expected to occur in the range of 100-150 cm^{-1}, while "Poley absorption" in the spectra of liquids and polymers usually appears at frequencies of 10-100 cm^{-1}, and a qualitative assessment of the Poley absorption frequency for Vectra A 950 according to the usual scheme [136] gives the value $\nu_{libr} = 65\ cm^{-1}$. This circumstance, since the intensity of the band at 127 cm^{-} is rather significant, leads authors who have studied the FIR spectra of mesomorphic substances [136, 248], to the conclusion about the specificity of librational motion in the liquid crystal. They believe that the difference between the calculated value of the libration frequency and the frequency of the experiment is a consequence of a deeper and narrower energy well in which mesogens librate. The LC medium provides a steeper potential well with respect to angular displacements around the long axis of the mesogen (i.e., with a lower moment of inertia and less hindered rotation) and, therefore, a higher value librational frequency.

Such a detailed discussion of the assignment of absorption at 127 cm^{-1} in the FIR spectrum of Vectra A 950 was necessary because it was this band that turned out to be the most sensitive to orientational ordering of the copolymer. The dichroic ratios of most of the bands in the FIR spectrum (see Table 6) and bands in the mid-IR region (see Figure 60) for granules and fibers are approximately the same $R_g \approx R_f$, and if they differ, then insignificantly, within the measurement error. For example, the dichroic ratio of the stretching vibration band of the carbonyl group at 1727 cm^{-1} in the spectrum of granules is $R_g = 0.85$, and in the spectrum of fibers, $R_f = 1.3$, while the analogous values for the band at 127 cm^{-1} in the FIR spectrum differ several times: $R_g = 0.45$ and $R_f = 1.55$.

As a result, the parameters of the orientational order obtained from the stretching vibration band for granules and fibers (mid-IR range) practically coincide, which does not allow us to confidently speak of different orientations of chain fragments in granules and fibers. Whereas the parameters of the orientational order obtained from the band of torsional vibrations at 127 cm^{-1} are different for granules and fibers. Their values are equal to $S_g = -0.22$, $S_f = 0.15$, which makes it possible to distinguish more definitely the orientation of chains in fibers and in granules.

Unlike the LC copolymer Vectra A 950, Armos is a rigid-chain polymer, the macromolecules of which retain the high orientation typical of LCP. Figure 49 shows the IR spectra of its fibers in the range of 4000-400 cm^{-1} before and after heat treatment. The assignment of bands in the IR spectra of Armos is presented in [251, 252].

Heat treatment, as can be seen from Figure 62, leads to a significant change in the IR spectrum. Especially at the frequencies of manifestation of stretching vibrations in NH groups (3700 - 2500 cm^{-1}), hydrogen bonded (HB). Redistribution in intensity and shift of absorption bands is also observed at frequencies of manifestation of stretching vibrations in C = O groups and at frequencies of bending vibrations in NH groups in the region of 1700 - 500 cm^{-1}. Such changes indicate not only the transformation of a system of hydrogen bonds in polymer macromolecules during heat treatment, but also about the structural rearrangement of the polymer chains themselves. So, growth is intensive the low-frequency wing of the anomalously wide band in the range of 3300 - 2800 cm^{-1} is associated [253, 254] with the enhancement of intramolecular HB in benzimidazole and amide fragments.

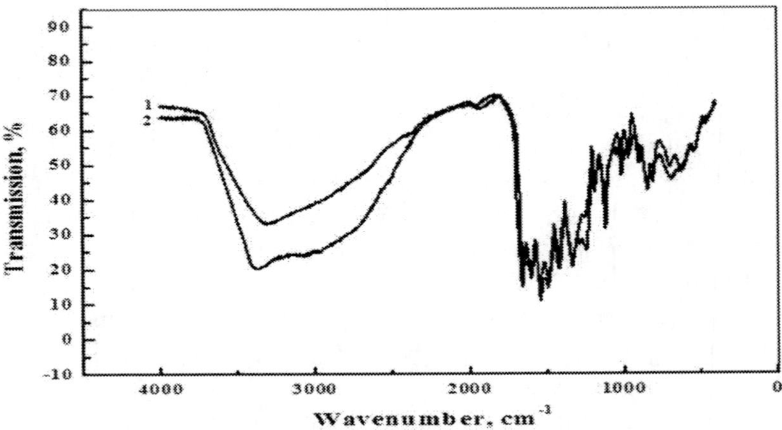

Figure 62. IR spectra of Armos. Initial fibers-1, heat-treated-2.

The fact that strong HBs between benzimidazole fragments play an important role in the formation of intermolecular associates in the polymer chains of Armos is also evidenced by changes in the region of 1700 - 500 cm^{-1}. Here, a redistribution of intensity is observed in the band at 1664 cm^{-1} (the so-called amide 1) in favor of its high-frequency component, and also in the doublet at 1324 - 1250 cm^{-1} (the so-called amide 3). The band at ~ 690 cm^{-1} (NH bending vibration) exhibits its structure, implying the appearance of ordered regions in the polymer after heat treatment.

Additional information, regarding the changes caused by the heat treatment of Armos is carried by FIR spectra in the range of 20 - 500 cm^{-1}, in which torsional (torsional) vibrations on the C - O, C - C, C - N, and N - H bonds, as well as librational vibrations of benzene and phenylbenzimidazole rings, are manifested, that is, the dynamics of larger-scale groupings than interatomic vibrations. In Figure 63 shows these spectra for the original and heat-treated Armos fibers. It is seen that the FIR spectrum heat treated Armos fibers differ significantly from the spectrum of the original fibers, especially at the lowest frequencies.

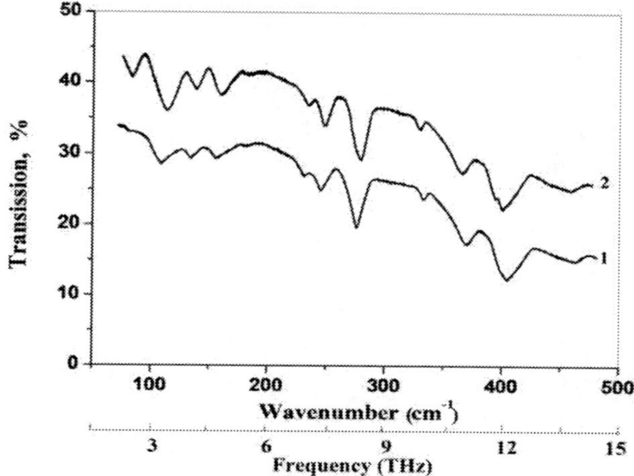

Figure 63. FIR-spectra of Armos. Initial fibers-1, heat-treated-2. Spectrum 2 is shifted along the ordinate.

Table 7. Spectral characteristics of the DIR absorption spectra of the Armos

Vibrational modes RCP Armos, (cm^{-1})	Approximate description	Molecular dynamics
80	τ CO(69)+τ (CC)(10)+ tr.	Lattice vibrations
111	τ CCO(28)+ τ B ring(12)+ libr.	
134	δ COC(46)+ δ CC B ring (10)+ libr.	
157	δ CCO(20)+ τ BIZ ring(34)+ libr.	
228	τ B ring(57)+ τ CC(15)	Torsion vibrations
244	τ NH(38)+ δ CC(32)+ δ B ring(28)	
275	τ BIZ ring (45)+τ CONH(28)	
329	δ CN(39)+γ CO(13)	Deformational vibrations
364	δ BIZ ring (24)+ δ B ring(14)+ γ CC(14)	
401	δ B ring(74)+ γ B ring(14)	
475	γ CCC(36)+ δ COC(43)	

Symbols: tr.-translation, libr.-libration, τ-torsion, δ- and γ-bending vibrations, B ring-benzene ring, BIZ ring- benzimidazole ring.

To identify absorption bands in the FIR spectrum of Armos, we used experimental and calculated literature data on the terahertz IR spectra of aminophenyl benzimidazole [255, 256] and phenylene terephthalamide (PPTA) [257], which are part of the monomer unit of this polymer. Based on these data, the preliminary assignment of bands in the FIR spectrum of Armos fibers is presented in Table 7.

Absorption from the natural vibrations of the HB lies in the same range and manifests itself in the FIR spectrum of Armos by the background, half-width and shift of the maximum of the bands, since the vibrations of the HB in ordered structures are not localized on separate groups, but are rotational or translational vibrations.

The performed assignment means that the entire FIR spectrum of Armos is in the range of 50-450 cm^{-1} actually characterizes the features of the limited small-angle motion of fragments of macromolecules of this polymer.

As in the case of the LCP of Vectra A950, the FIR spectrum of Armos fibers changed most noticeably after their heat treatment at the frequencies of libration of mesogenic groups. True, in contrast to the LCP, the absorption band by the Poley mechanism in the Armos spectrum more clearly manifested its fine structure of lattice vibrations, indicating a greater crystal-like ordering of the structure of the heat-treated polymer.

Another spectral effect is the high-frequency shift of the torsional vibration bands in the range of 200-300 cm^{-1}. The bands of torsional vibrations between benzene rings and amide groups at 228 and 244 cm^{-1} are shifted by 3-4 cm^{-1}, and the bands of torsional vibrations between phenylamide segments and benzimidazole rings at 275 cm^{-1} are shifted by ~ 4 cm^{-1}. These changes in the spectrum also indicate the hardening of Armos polymer chains during heat treatment.

All these changes in small-angle molecular dynamics undoubtedly contributed to the redistribution of hydrogen bonds in Armos macromolecules during HT, which is recorded from the mid-IR spectra (see Figure 60). However, for the structural modification of Armos during thermal annealing, one cannot do without large-scale motion of the chains. Such, called by the authors "quasi-segmental ", the thermal motion of the chains was studied in detail in a series of works [230-243, 258] by the method of proton magnetic resonance (PMR) of wide lines.

These studies have shown that during heat treatment, structural rearrangements of the "healing" type and orientation of the chains occur, and the possibility of spontaneous self-organization appears due to the movement of fragments of rigid LC chains relative to each other. Such quasi-segmental (in contrast to the segmental movement of flexible chains) mobility occurs under conditions of a high degree of straightening and mutual orientation of macromolecules in the melt and conformational movement in them is the result of random accumulation of displacements arising from local vibrations of chain links, i.e., according to the Bresler-Frenkel mechanism [259]. The possibility of experimental investigation of local forms of motion (libration of mesogens and torsional vibrations around fixed bonds) in LC (Vectra) and LC (Armos) polymers from their FIR spectra is shown above.

In general, the features of molecular dynamics revealed in a comparative analysis of LC and LC polymers by their IR and PMR spectra make it possible to more clearly represent the picture of the structural modification and dynamics of mesophases at the molecular level during thermal phase and relaxation transitions in such systems.

Conclusion

Although the share of FIR spectra in extensive studies using molecular spectroscopy is relatively small, its use has opened up fundamentally new possibilities for studying and characterizing polymers that are not covered by the mid-IR region. The data presented in the review provide numerous examples of such kind.

In contrast to the traditional applications for IR spectroscopy, based on the study of local stretching and deformation vibrations of atoms, which are relatively weakly dependent on intermolecular interactions, the FIR range carries new information about the torsional vibrations of fragments of macromolecules of various sizes, which, as it turned out, were not controlled in many ways. low intramolecular force constants, and potential barriers to intermolecular interactions.

The problems of lattice and quasilattice vibrations, sensitivity to the conformational state of the chain and anharmonicity of vibrations, natural vibrations of hydrogen bonds, motion of heavy ions in ionomers and Coulomb interactions are considered in detail in this review; and they all remain attractive for future research.

For studying the dynamics of polymers, the low-frequency infrared region is of decisive interest also because in this range of the electromagnetic spectrum the resonance type of the spectrum coexists with the relaxation one. Therefore, special attention in the review was paid to obtaining information on molecular motions leading to δ-, γ-, and β-transitions in glassy polymers. Experimental confirmation of the mechanism of δ-relaxation was obtained as limited rotation (libration) of monomer units, as well as β-relaxation as a reorientation movement of a chain section close to the value of the statistical segment with overcoming, first of all, the intermolecular potential barrier. It was found that the main characteristics of these transitions are related to the main molecular parameters of polymers, such as the structure of the monomer unit, thermodynamic chain stiffness, cohesion energy, and potential barrier internal rotation.

The limited scope of this review does not allow us to consider many interesting articles on the low-frequency IR spectra of polymers. These works include studies of PE [260-263], PP [264, 265], PET [266], PTFE [267], polyesters [268, 269], polyamides [270, 271], PMMA [271, 273],

PVF2 [274, 275], TPX [276], PS [277], polyacetylene [278, 279], poly (N-vinylcarbazole) [280], copolymers [281], polypeptides, proteins [282], liquid crystal polymers [283, 284] and BC in polymers [285], as well as POE [286], methacrylates [287, 296, 297], industrial polymers [288-290], and PA-6 [291].

The range of issues of modern physics of macromolecules touched upon in this review indicates the great possibilities of low-frequency IR spectroscopy for studying the chemical and physical structure, as well as the molecular dynamics of polymeric materials. We are confident that many more applications of it will appear in the future as instrumental methods improve and theoretical understanding grows. Just as it happened at the beginning of the century in connection with the advances in terahertz spectroscopy.

References

[1] Amrhein EM, Muller FH (1969) Kolloid Z.Z. *Polym.* 234: 1078.

[2] Chantry GW, Chemberlain JE (1972) in: Jenkins AD (ed.) *Polymer science. A materials science handbook.* NorthHolland, Amsterdam, p. 1373..

[3] Hummel DO (1966) I*nfrared spectra of polymers in the medium and long wavelength regions.* Interscience Publ., New York London Sydney.

[4] Frank WFX and Leute U (1983) in: Button KL (ed) I*nfrared and millimeter waves,* vol. 8, p. 51. Academic Press Inc., New York.

[5] Krimm S (1960) *Adv. Polym. Sci.* 2:51.

[6] Tasumi M, Krimm S (1967) *J. Chem. Phys.* 46: 755.

[7] Mushrooms LA (1977). *Theory of infrared spectra of polymers.* “Science”, Moscow, p.163.

[8] Zbinden R (1964) *Infrared spectroscopy of high polymers.* Academic Press, New YorkLondon.

[9] Krimm S (1963) in: Davies M (ed.) *Infrared spectroscopy and molecular structure.* Elsevier, Amsterdam.

[10] Zerbi G (1975) in: Califano S (ed.) *Lattice dynamics and inter molecular forces.* Academic Press, New York, p. 384.

[11] Kumpanenko IV, Tshuhanov NV (1981) *Uspehi khimii* 50:1627.

[12] Ozerkovskii BV, Roshchupkin VP (1980) *Dokl Akad Nauk SSSR* 254:15712..

[13] Frank WFX and Rabus G (1974) *Colloid Polym.Sci.* 252:1003.

[14] Moller KD, Rothschild WG (1971) *Far- infrared spectroscopy.* Willey, New York, p.495.

[15] Frank WFX (1979) in: Reed RP and Hartwig G (eds.) Nonmetallic materials and composites at low temperatures. *Plenum,* New York,p 51.

[16] Leute U, Grossman HP (1981) *Polymer* 22:1333.

[17] Fleming JM, Chantry GW, Turner PA, Nicol EA, Willis HA, Cudby MEA (1972) *Chem. Phys. Lett.* 72:84.

[18] Bur AJ (1985) *Polymer* 26:963.

[19] Amrhein EM (1972) *Ann. N.Y. Acad. Sci.* 196:179.

[20] Goldstein M, Seeley ME, Willis HA, Zichy VJI (1973) *Polymer* 14:530.

[21] Dechant J (1972) *Ultrarotspektroskopische untersuchungen an Polymeren.* Akademie Verlag, Berlin.

[22] Tadokoro H (1979) *Structure of crystalline polymers.* Willey, New York.

[23] Amrhein EM, Frischkorn H (1973) *Kolloid Z.Z.Polym.* 251:369.

[24] Hannon MJ, Boerio FJ and Koenig JL (1969) *J. Chem. Phys.* 50:2829.

[25] Chantry GW, Fleming JW, Nicol EA, Willis MA, Cudby MEA, Boerio FJ (1974) *Polymer* 15:69.

[26] Chantry GW, Fleming JW, Nicol EA, Willis HA, Cudby MEA, Boerio FJ (1977) *Polymer* 18:37.

[27] Manley TR, Martin CG (1971) *Polymer* 12:524.

[28] Krimm S, Folt VL, Shipmen JJ, Berens A (1963) *J. Polym. Sci.,* Polym. Phys. Ed. 1:2621.

[29] Jasse B, Monnerie L (1975) *J. Phys.* D 8:863.

[30] Bershtein VA, Ryzhov VA (1982) *Fizika Tv Tela* 24:162.

[31] Bershtein VA, Ryzhov VA (1984) *J Macromol Sci - Phys* B 23:271.

[32] Ryzhov VA, Tonkov MV (1973) In: Denisov GS (ed) *Molecular spectroscopy,* 2 ser. (in Russian). University Press, Leningrad.

[33] Ryzhov VA, Bershtein VA (1987) *Vysokomol Soedin* A 29:1852.

[34] Neppel A, Butter IS (1984) *Spectrochim. Acta* A 40:1095.

[35] Wright CJ (1976) in: Ivin KJ (ed.) *Structural studies of macromolecules by spectroscopic methods.* Willey, London.

[36] Spell SJ, Shepherd IW, Wright CJ (1977) *Polymer* 18:905.

[37] Hoffman V, Frank W, Zeil W (1970) *Kolloid Z.Z.Polym.* 241:1044.

[38] Belopolskaya TV (1968) *Dokl Akad Nauk SSSR* 180:1588.

[39] Berhstein VA, Egorov BM, Ryzhov VA (1986) *Vysokomol Soedin* B 28:268.

[40] Chantry GW, Fleming JW, Cook RJ, Moss DG, Nicol EA, Willis NA, Cudby MEA (1973) *Infrared Phys.* 13:157.

[41] Liang CJ, Krimm S (1958) *J. Polym. Sci.* 31:513.

[42] Godovskii YuC (1976) *Teplofizicheskie metody issledovaniya potimerov.* Ed "Khimiya", Moscow.

[43] Bershtein VA, Emelyanov YuA, Stepanov VA (1984) *Vysokomol. Soedin.* A 26:2272.

[44] Bershtein VA, Ryzhov VA (1982) *Vysokomol. Soedin* B 24:495.

[45] Ozerkovskii BV, Roshchupkin VP (1979) *Dokl Akad Nauk* SSSR 248:659.

[46] Bershtein VA, Egorov VM, Ryzhov VA, Sinani AB, Stepanov VA (1981) *Fizika Tv Tela* 23:1611.

[47] Zerbi G, Piseri L, and Cabassi F (1971) Molec. Phys. 22:241.

[48] Sverdlov GW, Kovner MA, Krainov EP (1970) *Kolebatelnye spektry mnogoatomnykh molecul.* Ed. Nauka, Moscow48..

[49] Shimanouchi T, Koyama J, Itoh K (1974) *Progr. Polym. Sci.* Japan 7:273.

[50] Chantry GW, Evans MW, Chamberlin J and Gebbie HA (1969) *Infrared Phys.* 9:85.

[51] Davis M, Pardoe GWF, Chemberlin JE, Gebbie HA (1970) *Trans. Farad. Soc.* II 66:273.

[52] Brot C, Darmon I (1971) *Molec.Phys.* 21:785.

[53] Jain SR, Walker S (1971) *J.Chem.Phys.* 75: 2942.

[54] Chantry G (1983) in: Button KL (ed.) *Infrared and millimeter waves,* vol. 8, ch.1. Academic Press, New York.

[55] Chantry GW (1967) *Nature* 214: 163.

[56] Chantry GW, Gebbie HA (1965) *Nature* 208: 378.

[57] Larkin I (1973) *Trans.Faraday Soc.*II 69: 1278.

[58] Hill N, Vangan W, Price A, Davies M (1969) *Dielectric properties and molecular behaviour.* Van Nostrand, Princeton, New York.

[59] Higasi K, Minami R, Takahashi H, Ohio A (1973) *Trans. Faraday Soc.* II 69:1579.

[60] Reid CJ, Evans MW (1982) *J. Chem. Phys.* 76: 2576.
[61] Chantry GW, Fleming JW, Nicol EA (1972) *Infrared Phys.* 12:101.
[62] Johnson KW, Rabolt JF (1973) *J. Chem. Phys.* 58: 4536.
[63] Piseri L, Powell BM, Dalling G (1973) *J.Chem.Phys.* 58:158.
[64] Viras F, King TA (1984) *Polymer* 25:899.
[65] Viras.F, King TA (1990) *J. Non-Crystalline Solids* 119: 65.
[66] Wendorff J (1977) in: *Proc. 4th Internat. Conf. Phys. Non-Cryst. Solids.* Aedermansdorf, p.94.
[67] Warrier AVR and Krimm S (1970) *Macromolecules* 3: 709.
[68] Tasumi M, Shimanouchi T (1971) *Polymer J.* 2:62.
[69] Moore WH, Krimm S (1973) *J. Chem. Phys.* 59:5195.
[70] Moore WH, Krimm S (1975) Makromol. *Chem. Supplm.* 1:491.
[71] Goldstein M, Stephenson D, Maddamas WF (1983) *Polymer* 24:823.
[72] Ryzhov VA, Bershtein VA (1989) *Vysokomol Soedin* A 31:451.
[73] Bershtein VA, Ryzhov VA (1985) *Dokl Akad Nauk SSSR* 284:890.
[74] Evans MW, Evans GJ, Coffey WT, Gricolini P (1982) *Molecular dynamics and theory of broad band spectroscopy.* Wiley Interscience Publ., New York, Chichester, Brisbane, Toronto, Singapore.
[75] McLelan CK, Walker S (1977) *Canad. J. Chem.* 55: 2411.
[76] Pardoe GWF (1970) *Trans. Faraday Soc.* II 66:2699.
[77] Bershtein VA, Ryzhov VA, Ganicheva CH, Ginzburg LI (1983) *Vysokomol Soedin* A 25:1385.
[78] North AM (1975) *J. Polym. Sci., Polym. Symp.* 50:345.
[79] Bossis G (1982) *Physica* A 110:408.
[80] Jakobsen RJ, Brasch JW (1964) *J. Amer. Chem. Soc.* 86:3371.
[81] Knozinger E, Jacox M (1978) *Ber. Bunsenges. Phys. Chem.* 88:57.
[82] Ryzhov VA, Bershtein VA (1988) In: Tudos F (ed) *8th Europ.Symp. Polym.Spectr.* (ESOPS 8), 2326 Aug. 1988. Budapest, Hungary.
[83] Tonelli AE (1971) *Macromolecules* 6:682.
[84] Arning HJ, Dorfmtller Th (1984) *Infrared Phys.* 24:221.
[85] Bondi A (1968) *Physical properties of molecular crystals, liquids and glasses.* Willey, New York, p.29.
[86] Entsiklopediya polimerov, vol (1974) ed *Sov. entsiklopediya,* Moscow, p 726.
[87] Tager AA (1978) *Fiziko-khimiya polimerov* Ed. Khimiya, Moscow.
[88] Askadskii AA, Kolmakova LK, Tager AA, Slonimskii GL, Korshak VV (1977) *Vysokomol Soedin* A 19:1004.
[89] McCrum NG, Read BE, Williams G (1967) *Anelastic and dielectric effects in polymeric solids.* John Wiley and Sons, New York.
[90] Heijboer J (1977) *Intern. J.Polym.Mater.* 6:11.
[91] Johari G and Goldstein M (1973) *J.Chem.Phys.* 58:1766.
[92] Williams G (1977) *Dielectric and related molecular processes,* vol. 2 (Spec. Period. Report, the Chemical Society, London).
[93] Colin J, Reid CJ, Evans MW (1979) *Trans.Faraday Soc.* II 75:1218.
[94] Vij JK, (1983) *Nuovo Cimento* D 2:751.
[95] Ashcraft CS, Boyd RH (1976) *J. Polym. Sci.,* Polym. Phys. Ed. 14:2153.

[96] Buckingham KA and Belling JW (1981) *Proc. IEE* 128:215.
[97] McCammon RD, Saba RG, Work RN (1969) *J. Polym. Sci.,* Polym. Phys. Ed. 7:1721.
[98] North AM, Parker TG (1972) *Trans. Faraday Soc.* II 68:1.
[99] Amrhein EM (1967) *Kolloid Z. Z. Polym.* 216217:38.
[100] Poley JP (1955) *J. Applied Sci.* B 4:337.
[101] Lunkerheimer P, Schneider U, Brand R, Loidl A (2000) *Contemporary Physics* 41:15.
[102] Ngai KL, Paluch (2004) *J. Chem. Phys.* 120:857.
[103] Ngai KL (2003) *J. Phys. Condensed Matter* 15:S1107.
[104] Taraskin S, Simdyankin, Elliott S, Neilson J, Lo T (2006) *Phys. Rew. Letters* 97:1.
[105] Lunkerheimer P, Loidl A (2003) *Phys. Rew. Letters* 91:20.
[106] Ryzhov VA (2017) *Solid State Physics* 59: 1422.
[107] Jin Y-S, Kim G-J, Jeon S-G (2006) *J. Korean Phys. Soc.* 40:513.
[108] Caliskan G, Kisliuk A, Novikov VN, and Sokolov AP (2001) *J. Chem. Phys.* 114:10189.
[109] Kitai MS, Nazarov MM, Nedorezova PM, Shkurinov AP (2017) *Quant. Electr.* 60:409.
[110] Fedulova E, Nazarov M, Kitai MS, Shkurinov AP (2011) *Proceedings of SPIE* 8337:833701.
[111] Birshtein TM, Ptitsyn OB (1964) *Konformatsia makromolecul.* Ed. Nauka, Moscow.
[112] Volkenstein MV (1963) Conformational statistics of polymer chains. *Interscience,* New York.
[113] Kalmykov JuP, Gaiduk VI (1981) *J Fiz Khimii* 55:305111..
[114] Van Crevelen DW (1972) *Properties of polymers: Correlation with chemical structure.* Elsevier, Amsterdam and London.
[115] Miller S, Tomazawa M, McCrone RK (1972) *Molec. Cryst.* 7:54.
[116] Hayler L, Goldstein M (1977) *J. Chem. Phys.* 66:4336.
[117] Irvine JD, Work RN (1973) *J. Polym. Sci.,* Polym. Phys. Ed. 11:179.
[118] Hedvig P (1987) *Dielectric properties of polymers.* Wiley, New York.
[119] Sauer O (1971) *J. Polym. Sci.,* Polym. Symp. 2:69.
[120] Glasston S, Landler K, Ehring G (1948) *The theory of absolute reaction rates* Ed. "Science", Moscow.
[121] Johari GP (2002) *J. Non- Cryst. Solids* 307-310:114.
[122] Shuker R, Gammon RW (1970) *Phys. Rev. Lett.* 25:222.
[123] Bembenek SD, Laird BB (2001) *J. Chem. Phys.* 114:2340.
[124] Ryzhov VA (2003) *Physicochemistry of polymers.* Tver: Tver. state un-t. 9:37.
[125] Libov VS, Perova TC (1992) *Tr. GOI. SPb.* 81: 3.
[126] Larkin IW (1973) *J. Chem. Soc. Faraday Trans.* II 69:1278.
[127] Novikov VN, Sokolov AP, Stube B, Surovtsev NV, Duval E, Mermet A (1997) *J. Chem. Phys.*107:1057.
[128] Sokolov AP, Kisliuk A, Soltwisch M, Quitmann D (1992) *Phys.Rev. Lett.* 69:1540.

[129] Bershtein VA, Egorov VM (1990) *Differential scanning calorimetry in the physics and chemistry of polymers.* "Chemistry" Publ House, Leningrad [Translation: Differential Scanning Calorimetry of Polymers, Ellis Horwood~ Chichester, 1993].

[130] Buchenau U, Pecharreman C, Zorn R, Frick B (1996) *Phys. Rev. Lett.* 77:659.

[131] Bergman R, Svanberg C, Andersson D, Brodin A, Torell LM (1998) *J. Non-Cryst. Solids* 235–237:225..

[132] Gottlieb YY, Darinsky AA, Svetlov YE (1986) *Physical kinetics of Macromolecules.* Ed. "Chemistry", Leningrad, c. 272..

[133] Surovtsev NV, Wiedersich JA, Novikov VN, Rossler E, Sokolov AP (1998) *Phys. Rev.* B58:14888.

[134] Terao W, Mori T, Fujii Y, Kojima S (2017) *Spectrochimica Acta Part A Molecular and Biomolecular Spectroscopy* 192:.

[135] Kabeya M, Mori T, Fujii Y, Koreeda A, Lee BW, Ko JH, Kojima S (2016) *Phys. Rev.* B94:224204.

[136] Bershtein VA, Ryzhov VA (1994) *Adv. Polym. Sci.* 114:43.

[137] Ryzhov VA (2002) *Physics of the Solid State* 44: 2229.

[138] G¨otze W, Sjogren L (1992) *Rep. Prog. Phys.* 55:241.

[139] Kanaya T, Kaji K (2001) *Adv. Polym. Sci.* 154:87.

[140] Gilroy KS, Phillips WA (1981) *Phil. Mag.* 43:735.

[141] Surovtsev NV, Achibat T, Duval E, Mermet A, Novikov VN (1997) *J. Phys: Condens. Matter.* 7:8077.

[142] Cummins HZ, Li G, Hwang HY, Shen GQ, Du WM, Hernandez J, *Tao NJ* (1997) Z. Phys. B 103:501.

[143] Zhang C, Arrighi V, Gagliardi S, McEwen IJ, Tanchawanich J, Telling MT, Zanotti J-M (2006) *Chem. Phys.* 328:53.

[144] Cowie JWG (1980) *J. Macromol. Sci., Phys.* B 18: 568.

[145] Bershtein VA, Ryzhov VA (1982) in: Lindberg J (ed) *6th Europ. Symp. Polym. Spectr.*(ESOPS 6), 1113 Aug. 1982. Hameenlinna, Aulanko.

[146] Bershtein VA, Egorov BM, Emelyanov YuA, Stepanov BA (1983) *Polymer Bull* 9:98.

[147] Bershtein VA, Egorov BM (1985) *Vysokomol Soedin* A 27:2440.

[148] Struik L (1976) *Ann. N.Y. Acad. Sci.* 279:78.

[149] Tsvetkov VN, Eskin VE, Frenkel SYa (1964) *Struktura makromolecul v rastvorah.* Ed. Nauka, Moscow.

[150] Gribov LA, Zubkova OB, Shabadash AN (1973) *Z Priki Spektr* 18:1028.

[151] Zerbi G (1977) in: Barnes AJ, OrvilleThomas WJ (eds) *Vibrational spectroscopy modern trends.* Elsevier, New York.

[152] Smirnov BR, Plotnikov VD, Ozerkovskii BV, Roshchupkin VP, Enikolopian NS (1981) *Vysokomol Soedin* A 23:2588.

[153] Ryzhov VA, Bershtein VA (1989) *Vysokomol Soedin* A 31:458.

[154] Bershtein VA, Pertsev NA (1984) *Acta Polym* 35:575.

[155] Bershtein VA, Egorov VM, Emelyanov YuA (1985) *Vysokomol Soedin* A 27:2451.

[156] Belopolskaya TV (1969) *Opt Spectr* 26:476.

[157] O Reilly J (1977) *J. Appl. Phys.* 48:4043.
[158] Egorov EA, Zhizhenkov VV (1982) *J. Polym. Sci., Polym. Phys.* Ed. 20:1089.
[159] Gibbs J, DiMarzio E (1957) *J. Chem. Phys.* 28:373.
[160] Adams G, Gibbs J (1965) *J. Chem. Phys.* 43:139.
[161] Bresler SE, Ierysalimskii TV (1965) *Fizika i khimiya makromolecul.* Ed. Nauka, Moscow- Leningrad, p 81.
[162] Heijboer J (1965) in: Heijboer J (ed.) *Physics of noncrystalline solids.* Holland Publ.Comp., Amsterdam, p.231.
[163] Boyer R (1976) *Polymer* 17:996.
[164] Bartenev GM (1979) *Struktura i relaksatsionnyia svoistva elastomerov.* Izd. "Khimiya", Moscow.
[165] Kolaric J (1982) *Adv. Polym. Sci.* 46:119.
[166] Hayakawa K, Wada J (1974) *J. Polym. Sci.,* Polym. Phys. Ed. 12:2119.
[167] Ryzhov VA, Bershtein VA, Sinani AB (1990) *Vysokomol.Soedin.* A 32:90.
[168] Aras L, Baysal BM (1964) *J. Polym. Sci.,* Polym. Phys. Ed. 22:1453.
[169] Baccaradda M, Butti E, Frosini V (1966) *J.Appl.Polym.Sci.* 10:399.
[170] Bershtein VA, Egorov VM, Podolsky AF, Stepanov VA (1985) *J. Polym. Sci.,* Polym. Lett. Ed. 23:371.
[171] Bershtein VA, Emelyanov YuA, Stepanov VA (1981) *Mehan. Komposit.Mater.* 1:9.
[172] Desando MA, Rashem M, Siddiqui NA, Walker S (1984) *Trans. Faraday Soc.* II 80:747.
[173] Finch A, Gates P, Radcliffe K, Dickson F, Bentley F (1970) *Chemical application of far infrared spectroscopy. Academic* Press, London, New York.
[174] Jakobsen R, Brasch JW (1965) *Spectrochim. Acta* 20:1753.
[175] Frank WFX, Fiedler H (1979) *Infrared Phys.* 19:481.
[176] Jakes J, Krimm S (1971) *Spectrochim. Acta* 27:35.
[177] Beiopolskaya TV (1976) *Vestnik LGU* 22:44.
[178] Bessler E, Bier G (1969) *Makromol. Chem.* 20:122.
[179] Tadokoro H, Kobayashi M, Yoshidome H, Tai K and Makino D (1968) *J. Chem. Phys.* 49:3359.
[180] Shen DY, Pollack SK, Hsu SL (1989) Macromolecules 22:2564.
[181] Feairheller WR, Kotton JE (1967) *Spectrochim. Acta* 23:2225.
[182] Stanevich AE (1964) *Opt. Spectr.* 16:243.
[183] Jakobsen RJ, Mikawa J, Brasch JW (1967) *Spectrochim.Acta* 23:2199.
[184] Belopolskaya TV, Trapeznikova ON (1966) *Opt. Spectr.* 20:246.
[185] Ryzhov VA, Bershtein VA (1992) in: Skorohodov S (ed) *10th Europ.Symp. Polym.Spectr.* (ESOPS 10), 29 sept.2 oct. St.Petersburg, Russia.
[186] Holliday L (1975) Ionic polymers. Applied Science, London, p.35.
[187] Eisenberg A and King M (1977) *Ioncontaining polymers: Physical Properties and Structure.* Academic Press, New York, p.141.
[188] Eisenberg A (1970) *Macromolecules* 3: 147.
[189] Neppel A, Bulter IS, Eisenberg A (1979) *Can.J.Chem.* 57:2518.
[190] Rouse GB, Risen WM, Tsatsas AT, Eisenberg A (1979) *J. Polym. Sci.,* Polym. Phys.Ed. 17:81.

[191] Suchocka-Galas K (1987) *Europ. Polym. J.* 12:951.

[192] Tsatsas AT, Reed JW, Risen WM (1971) *J. Chem. Phys.* 55:3260.

[193] Mattera VD, Risen WM (1984) *J. Polym. Sci.,* Polym. Phys. Ed. 22:67.

[194] Exarhos GJ, Miller PJ, Risen WM (1974) *J. Chem. Phys.* 60:4145.

[195] Viras F, King TA (1984) *Polymer* 25:1411.

[196] Tsatsas AT, Risen WM (1970) *J.Am.Chem.Soc.* 92:1789.

[197] Neppel A, Bulter IS, Eisenberg A (1979) *Macromolecules* 12:948.

[198] Slisenko OV, Lebedeva EV, Ryzhov VA, Grigorieva OP (2004) *X Ukrainian Conference on Polymers.* Kiev, Ukraine.

[199] Grossmann HP and Frank WFX (1977) *Polymer* 18:341.

[200] Frank WFX, Schmidt H, Wulf W (1977) *J. Polym. Sci., Polym. Symp.* 61:317.

[201] Latour M, Abo Dorra H, Galigne JL (1984) *J. Polym. Sci., Polym. Phys.* Ed. 22:345.

[202] Lu FJ, Waldman DA, and Hsu SL (1984) J.Polym. Sci., Polym. Phys. Ed. 22: 827.

[203] Yaroslavsky NG, Konovalov LV (1967) *Opt. and Spectr.* Ed. "Science", Leningrad Collection III: 283.

[204] Yaroslavsky NG, Zheludov BA, Stanevich AE (1956) *Opt. and Spectr.* 1: 507.

[205] Bank AS, Krimm S (1968) *J.Appl.Phys.* 39:4951.

[206] Frank WFX (1976): *Proc. 2nd Int.Conf.and Winter School on Submillimeter Waves,* San Pierto Rico, p.111.

[207] Willis HA, Cudby MEA (1976) in: Ivin KJ (ed) *Structural studies of macromolecules by spectroscopic methods.* Wiley Interscience Publ., New York, Sydney, Toronto.

[208] Willis HA, Cudby MEA (1975) *Polymer* 16: 74.

[209] Manley TR, Williams DA, (1969) *Polymer* 10:339.

[210] Bruno F (1980) *Ann.Rev.Phys.Chem.* 31:265.

[211] Siesler HW, HollandMoritz K (1980) *Infrared and Raman Spectroscopy of polymers.* Marcel Dekker Inc., New York, Basel, p.320.

[212] Roshchupkin VP, Andreev IS, Roshchupkina TA, Chplakh.n GM, Starobunskii AB (1976) *Uch.Zap.* Univ of Erevan 2:46.

[213] Roshchupkin VP, Lubovskii AB, Kochervinskii VV, Roshchupkina TA (1969) *Vysokomol. Soedin.* A 11:2505.

[214] Bosomworth DR, Gush HP (1965) *Canad. J. Phys.* 43:751.

[215] Tonkov MV (1970) in: Bulanin MO (ed.) *Spectroscopy of interactions of molecules.* University of Leningrad Press, Leningrad.

[216] Libov VS, Perova GS (1976) *Opt. mehan. prom.* 5:52.

[217] Feairheller WR, Willler JI (1971) *Appl.Spectr*. 25:175.

[218] Yamamoto K, Tominaga K, Sasakawa H, Tamura A, Murakami H (2005) *Biophysics J.* 89:L22.

[219] Xie A, He Q, Miller L, Salavi B, Chance MR (1999) *Biopolymers* 49:591.

[220] Itoh K, Shimanouchi T (1970) *Biopolymers* 9:383.

[221] Shen S, Santo L, Genzel L (1981) *Canad. J. Spectr.* 23:126.

[222] Shen S, Santo L, Genzel L (2007) *Intern. J. Infrared and Millimeter Waves* 28:595.

[223] Husain SR, Hasted JB, Rosen D, Nicol E, Birsh JR (1984) *Infrared Phys.* 24:209.

[224] Belmont ML, Ambarek AN, Sik V (1984) *Infrared Phys.* 24:215.
[225] Itoh K, Hinomoto H and Shimanouchi T (1974) *Biopolymers* 13:307.
[226] Jiang L, Li M, Li C, San B, Yu L, Jin B, Lin Y (2014) *J. Inter. Millimeter and Teraherz Waves* 35:871.
[227] Viera FS, Pasquini C (2014) *Annal. Chem.* 86:3780.
[228] Bulkin BJ (1976) *Adv. in Liqud Crystal* 2: 199.
[229] Czamecki WA, Okretic S, Siesler HW (1997) *J. Phys. Chem.* B101:374.
[230] Fontana MP, Anachkova E, Monda O (1994) *Molec Cryst. and Liquid Cryst.* 243:31.
[231] Seci T (2014) *Macromol. Rapid Commun.* 35:271.
[232] Rutz F, Hasek T, Koch M, Richter H, Ewert U (2006*) Appl. Phys. Letters* 89:221911.
[233] Azeyanangi, Kaneco T, Ohki Y (2018) Jap. J. of Appl. Phys. 87:050302.
[234] de Gennes P (1977) Physics of Liquid Crystals. Ed. "Mir", Moscow.
[235] Libov VS, Perova TS (1979) Opt. fur. prom. 5:52.
[236] Venugopolan S, Prasad S (1989) J. Chem. Phys. 72:4153.
[237] Kirov N, Simova P (1984) Vibrational Spectroscopy of Liquid Crystals. Publ. B.A.N., Sofia.
[238] Volokhina AV, Kudryavtsev GI (1988) in collection. *Liquid crystal polymers.* Ed. "Chemistry", Moscow.
[239] Levchenko AA, Antipov EV, Plate NA (1999) *Macromol. Symp.* 146:142.
[240] Egorov EA, Zhizhenkov VV (1999) *Polymer* 40:3891.
[241] Egorov EA, Zhizhenkov VV (2002) *Vysokomol. Connect.* 44A: 1119.
[242] Egorov EA, Zhizhenkov VV (2004) *Vysokomol. Connect.* 46V: 1965.
[243] Egorov EA, Zhizhenkov VV (2005) *Solid State Physics* 47: 914.
[244] Wiberg G, Hillborg H, Gedde UW (1998) *Polym. Eng. Sci.* 38: 1278.
[245] Savitsky AA, Bilibin AYu, Gorshkova IA. (1992) *Vysokomol. Connect* 34: 143.
[246] Egorov EA, Zhizhenkov VV (1982) *J. Polym. Sci. Polym. Phys. Ed.* 20:1081.
[247] Chung T-S, Jin X (2000) *Polym. Eng. Sci.* 40:841.
[248] Evans M, Davies M, Larkin J (1973) *Chem. Soc.,* Faraday Trans. II 69:1011.
[249] Takahashi M, Kawazoe Y, Ishikawa Y, Ito H (2009) *Chem. Phys. Lett.* 479:211.
[250] Ponseca S, Estacio E, Murakami H, Sarukura N, Pobre R,Tominaga K, Nishizawa J (2008) *J. Phys.: Conf. Ser.* 112:042073C.
[251] Abbronin LA, Rakitina VA, Gribanov VA, Shablygin MV (2002) *Fibre Chem.* 34:140.
[252] Shebanov SM, Novikov IK, A.Pavlikov IA, Anan, in OB, Gerasimov IA (2016) *Fibre Chem.* 48:158.
[253] Shablygin MB, Slugin IV, Mamonova TS, Novikova LA (2009) *Fibre Chem.* 41:254.
[254] Bagaqutdinova SP, Vavilina TS, Komissarov SB, Sklyarova SG, Shablygin MV (2011) *Fibre Chem.* 43:174.
[255] Sudha S, Karabacak M, Kurt M, Cinar M, Sundaraganesan N (2011) *Spectrochem. Acta* 84 A:184.
[256] Shyma Mary Y, Jojo PJ, Yohannan Panicker C, ChristianVan Alsenoy, Sanaz Ataei, Ilkay Yildiz (2014) *Spectrochem. Acta* 122 A:499.

[257] Wong PTT, Garton A, Carlsson DJ, Wiles DM (1980) *J. Macromol. Sci. Part Phys.* 18:313.
[258] Zhizhenkov VV, Egorov EA, Kvachadze NG (2011) *Physicochemistry of Polymers* 17: 3.
[259] Bresler CE, Frenkel YA (1939) *JETF* 8: 1094.
[260] Leute U, Frank WFX (1980) *Infrared Phys.* 20:327.
[261] Frank WFX, Schmidt H, Heise B, Chantry GW, Nicol EA, Willis HA, (1981) *Polymer* 22:17.
[262] Schlotter NE, Rabolt JE (1984) *Macromolecules* 17:1581.
[263] Birch JR (1990) *Infrared Phys.* 30:195.
[264] Chantry GW, Fleming JW, Pardoe GWF, Reddish W and Willis HA (1971) *Infrared Phys.* 11:109.
[265] Beckett DR, Chalmers JM, Mackenzie MW, Willis HA, Edwards HGM, Lees JS, Lord DA (1985) *Europ. Polym. J.* 21:849.
[266] Frank W, Knaupp D (1975) *Ber.Bunsenges.Phys.Chem.* 79:1041.
[267] Willis HA, Reddish W, Buckingham KA, LlewellynJones DT, Knight KJ, Gebbie HA (1981) *Polymer* 22:20.
[268] Chantry GW, Fleming JW, Smith PM, Cudby M, Willis HA (1971) *Chem. Phys. Lett.* 10:473.
[269] Kobayashi M, Morishita H, Shimomura M (1989) *Macromolecules* 22:3726.
[270] Matsubara I, Magill JH (1973) *J. Polym .Sci.,* Polym. Phys. Ed. 11:1172.
[271] Nie CS, Kremer F, Poglitsch A, Bechtold (1985) *J. Polym. Sci., Polym. Phys. Ed.* 23:1247.
[272] Tadokoro H, Chatani J, Kusanagi N, Yokayama M (1970) *Macromolecules* 3:441.
[273] Manley TR (1976) in: Jones DW (Ed.) *Introduction to the spectroscopy of biological polymers.* Academic Press, New York, p.119.
[274] Ryzhov VA, Bershtein VA (2008) *Physics of the Solid State* 50: 1901.
[275] Yang DC, Tomas EL, (1984) *J. Mater. Sci.* 3:928.
[276] Kochervinskii VV, Glukhov VA, Romandin VF, Sokolov VG, Lokshin BV, (1988) *Vysokomol. Soedin.* A 30:1916.
[277] Birch JR, Nicol EA (1984) *Infrared Phys.* 24:573.
[278] Jano O, Wada J (1971) *J. Polym. Sci., Polym. Phys. Ed.* 9:669.
[279] Genzel L, Kremer F, Poglitsch A (1984) *Phys. Rev.* B29:4595.
[280] Bechtold G, Gensel L, Roth S (1985) *Solid St.Commun.* 53:1.
[281] Fujno M, Mikawa H, Yokoyama M (1984) *J.Noncrystalline solids* 64:163.
[282] Latour H, Moreira RL (1987) *J. Polym. Phys.,* Polym. Phys. Ed. 25:1913.
[283] Hayest W, Praff FL, Wong KS, Raneto K, Yoshino K (1985) *J. Phys.,* Solid St. Phys. 19:L555.
[284] Blumstein A (1985) *Polymer liquid crystals.* Academic Press, New York.
[285] Evans GJ, McCicki AS, Evans MW (1986) *J. Mol. Liqids* 32:149.
[286] Khoury Y, Hellwig P (2017) *Chem. Communication* 53: 8389.
[287] Seneviratne V, Furneaux JE, Frech R (2002) *Macromolecules* 35:6392.
[288] Yamada T, Kaij T, Aoki, Janda C, Mizuno M, Saito S at al.(2016) *Jap. J. Appl. Phys.* 55:03DC11.
[289] Angelo FD, Mics Z, Bonn M, Turhinovich D (2014) *Opt. Express* 22:12475.

[290] Cunningham PD, Velder NN, Valejo FA, Hayden LM, Polishak B, Zhou X-H (2011) *J. Appl. Phys.* 109:043525.

[291] Marlina D, Hoshina H, Ozaki Y, Sato H (2019) *Polymer* 181:121790.

[292] Wietzke S, Janden C, Reuter M, Ficsher BM, Koch M (2011*) J Mol. Struct.* 1006:41.

[293] Nagai N, Fukasawa R, (2004) *Chem.Phys.Lett.* 338:479.

[294] Naftaly M, Miller RE (2007) *Proc IEEE* 95: 1658.

[295] Balahrishnan J, Ficsher BM, Abbot D (2009) *Appl. Opt.* 48: 2262.

[296] Alfihed S, Bergen MH, Holzman JF, Foulds IG. (2018) *Polymer* 153: 325.

[297] Suzuki H, Ishii S, Sato H, Yamamoto S, Morisawa, YHoshina H (2013*) Chem. Phys. Letters* 575:36.

.

Appendix

Applications

1. The Relationship between the Molecular Dynamics of Amorphous Polystyrene and Its Modifications and Absorption Parameters of Poley-Type Absorption

In this study, we measured and analyzed the long-wave IR spectra of polystyrene and its modifications obtained by replacing hydrogen atoms in the benzene ring with methyl groups and/or chlorine and bromine atoms. All these spectra have an anomalously wide asymmetric absorption band with a maximum in the terager range from 40 to 80 cm^{-1}, depending on the nature of the substituent. It can be attributed to absorption associated with libration (rotational vibrations) of phenolic rings of monomer units, that is, absorption by the Poley mechanism. The relationship between the spectral parameters of this absorption and the molecular characteristics of the studied polymers makes it possible to directly analyze the role of the molecular structure and intermolecular forces in the dynamics of their macromolecules. It was found that the heights of potential barriers to libration of monomeric units of polystyrene and its modifications, estimated from the analysis of Poley absorption, are close to the activation energies of low-temperature δ-relaxation in these glassy polymers. Comparison of the libration barriers of monomeric units of PS macromolecules and its modifications with activation barriers local segmental mobility in the same PS confirms that the β-process as an elementary act of segmental dynamics is due to the correlated librational movement of the chain section, which is statistically independent of the neighboring sections. In this sense, the universal δ-process is a high-frequency precursor to the β-process in polymers like polystyrene.

Introduction

The influence of the features of the molecular structure and interactions of polymer molecules on their local and segmental dynamics still remains in the circle of theoretical and experimental studies [1, 2].

In recent years, significant advances in this direction have been achieved with the combined use of high-frequency dielectric and low-frequency vibrational (IR and Raman) spectroscopy due to their unique ability to directly analyze the features of a disordered condensed medium, determined by both individual and collective properties of molecules and the nature of intermolecular forces [3, 4].

The use of the submillimeter range in such studies is of particular interest, since here, the spectrum of the relaxation type and resonance.

Nance spectra tend to coexist, and low-frequency vibrational spectra are a direct source of information on molecular mobility that precedes relaxation dynamics in glassy polymers. We are talking about the manifestation of low-angle rotational vibrations in the long-wave IR (FIR) spectrum of liquids and disordered materials, including polymers [5-7]. Such small-angle rotational vibrations (librations) of phenolic rings in monomer units of polystyrene (PS), which are responsible for δ-relaxation in this polymer [8], appear in the FIR spectrum of PS with an abnormally wide absorption band with a maximum at 80 cm^{-1} [9].

This absorption (also called poly absorption) in the FIR spectra of polymers has been studied both theoretically and experimentally. It was shown that its intensity is proportional to μ^2/I, where μ and I are the dipole moment and the moment of inertia of the side group (phenolic ring) of the polymer, respectively. The position of the maximum of this absorption is proportional to $(V/2I)^{1/2}$, where V is the height of the potential barrier for libration, comparable in magnitude with the activation energy of the δ-transition in the PS [7]. The found relationship between the spectral absorption parameters of Poley with the molecular characteristics of polymers makes it possible to analyze the role of the molecular structure and intermolecular forces in the dynamics of macromolecules.

In this work, the found correlations were used to determine the potential barriers to torsional vibrations in macromolecules of polystyrene (PS) and its derivatives, which differ in the structure of the benzene ring. Carrying out research on a similar series of polymer objects, when given and controlled their molecular parameters vary in a controlled manner, allows to reveal the role of the electronic structure and steric factor in such elementary acts of segmental movement, as torsional (librational) vibrations of monomer units, preparing the relaxation mobility of macromolecules.

The results obtained were compared with the data of other techniques, and in particular, with the estimates and conclusions of work [10] performed on the same objects by DSC and UV spectroscopy.

Materials and Experimental Techniques

The objects of research were PS and its modifications, synthesized using a complex multistage procedure; this led to the creation of a set of polymers with a controlled change in the structure of the benzene ring. The structure of the aromatic ring was changed by substituting methyl groups (CH_3) and/or chlorine (Cl) and bromine (Br) atoms for hydrogen atoms in different positions on the ring with different numbers of substitutions and different combinations of substituents [11]. The list of samples and some of their thermophysical and energy characteristics are shown in Table 1. The molecular weights of the polymers were in the range (1 ÷ 2) 10^5. The molecular weights and van der Waals volumes of their monomer units are also shown in Table 1.

Far infrared (FIR) spectra were recorded on single-beam vacuum spectrometers with diffraction gratings: from 20 to 150 cm^{-1} on a device developed at Leningrad State University and modernized at the Physicotechnical Institute using an OAP-7 receiver and a new filtration system, and from 150 cm^{-1} to 450 cm^{-1} on the "Hitachi" FIS-21 spectrometer (Japan).

The resolution for a signal-to-noise ratio of about 100 was 1–2 cm^{-1}. The accuracy of determining the frequency of the band maximum is 0.5-1 cm^{-1}. The absorption coefficient k (ν) [2] $= ln\ (J_0/J)/t\text{-}t_0$, where J_0 and J are the transmission of samples with thicknesses t_0 and t, was measured with an error of 5% to 10%. All spectra were measured for samples obtained in the form of films with a thickness of 200 to 300 μm; measurements were carried out at room temperature.

The objects of study were PS and its modifications obtained by complex multistage synthesis [7, 8], which resulted in a set of polymers with controlled variation in the structure of the benzene ring. Structure the benzene ring was varied by replacing the hydrogen atom at different points of the ring with methyl groups (CH_3) and chlorine (Cl) and bromine (Br) atoms with different numbers of substitutions and different combinations of substituents. A list of samples and some of their thermophysical and energy characteristics are given in [6]. The molecular weights of the polymers were in the range (1-2) 10^5. The molecular weights and van der Waals volumes of their monomer units are shown in the table.

[2] α (ν) in the main text.

Table 1. Molecular characteristics of PS and its modifications

Polymer no.	Polymer	Van der Waals volume V, 10^{-24} cm^3	Molecular weight M, 10^{-27} kg	Radius of the Lateral group R, 10^{-10} m	Moment of inertia $I\mu = 0.4MR^2$, 10^{-47} kg m^2	Dipole moment μ, 3.33×10^{-30} C m	Cohesion energy E_{coh}, 10^{-19} J	$\mu^2/I\mu$, 10^{16} C^2 kg^{-1}
1	Polystyrene	85	134	2.8	420	0.18	0.46	0.85
2	Poly(2-methylstyrene)	104	160	2.96	560	0.39	0.58	3.02
3	Poly(4-methylstyrene)	103	160	2.95	560	0.36	0.56	2.56
4	Poly(2,4-dimethylstyrene)	115	182	3.07	690	0.37	0.66	2.22
5	Poly(2,5-dimethylstyrene)	115	182	3.07	690	0.32	0.58	1.65
6	Poly(4-chlorostyrene)	112	195	3.03	720	1.65	0.57	43
7	Poly(4-chloro-3-methylstyrene)	117	216	3.08	820	1.82	0.58	44
8	Poly(2-chloro-3,4-dimethylstyrene)	135	240	3.24	1000	1.70	0.69	31.5
9	Poly(2-chloro-3,5-dimethylstyrene)	135	240	3.24	1000	1.50	0.77	24.8
10	Poly(2,4,5-trimethylstyrene)	126	306	3.16	1230	1.70	0.67	25.7
11	Poly(4-bromo-2,5-dimethylstyrene)	138	314	3.26	1340	1.65	0.83	22.5
12	Poly(2,3-dichloro-4,5-dimethylstyrene)	150	300	3.35	1350	2.40	0.8	47

Note: The V and M values are taken from reference books [24, 29] or (as in the case of μ and E_{coh}) were evaluated according to the additive scheme for the contributions of functional groups.

Results and Its Discussion

Figure (a - l) shows the FIR spectra of PS and its modifications in the range 20–450 cm^{-1}. Absorption at these frequencies in the spectra of substituted benzenes, styrenes, and polystyrenes, as found [12, 13], is mainly due to deformation vibrations of benzene rings (modes 9b, 10b, and 16a) mixed in the polymer with deformation vibrations of the main chain, as well as torsional vibrations of monomer units and CH_3 group, if any. The assignment of low-frequency bands observed in the FIR spectrum of PS (Figure 1a) proposed in [14, 15] can be represented as follows: an intense band at ~ 405 cm^{-1} is an out-of-plane deformation vibration of the C – C bond of the aromatic ring (mode 16a); weak bands at 325 and 295 cm^{-1} - due to bending vibrations of C-C-C and C-C-H groups of the main chain. The most intense band in the FIR spectrum of PS with a maximum at 218 cm^{-1} has a complex contour consisting of at least two components at 218 and ~ 245 cm^{-1}. The corresponding pair of bands was observed for model compounds of PS [15] and, according to calculations, the band at 218 cm^{-1} was attributed to the in-plane vibrations of the C-H bonds of the aromatic ring (mode 9b), and its high-frequency component at 245 cm^{-1} was associated with skeletal vibrations of the main chain. The mode 10b of out-of-plane vibrations of the C-H bonds of the aromatic ring, predicted by calculations for substituted PSs in the range from 150 to 200 cm^{-1}, does not appear as a separate band in the FIR spectrum of PSs. And, finally, the low-frequency anomalously wide absorption band in the FIR spectrum of PS with a maximum at about 80 cm^{-1} is mainly due to torsional vibrations (libration) of the aromatic ring, ie, it refers to absorption by the Poley mechanism [16–18].

The above assignment of the observed bands in the FIR spectrum of PSs makes it possible to reliably identify the main absorption bands in the FIR spectra of modified polystyrenes.

From those presented in Figure (and Figure 2) of the spectra, it can be seen that when the hydrogen atoms in the benzene ring are replaced by Cl, Br, or CH_3 groups, the most significant change in the FIR spectrum profile occurs in the range of absorption by the Poley mechanism. And also in the frequency range 150–350 cm^{-1}, where vibration bands 9b and 10b are located, the position and intensity of which, as shown in [13, 15], strongly depends on the nature of the substituent and its place in the benzene ring. In addition, it should be taken into account that modes 9b and 10b are especially characterized by there is a connection with the deformation vibrations of the polymer chain, making them conformationally sensitive. Therefore, despite the fact that the assignment to modes 9b and 10b of bands at ~ 155 cm^{-1} and 375 cm^{-1} in the spectrum of poly (4-chlorostyrene) (Figure 1f) and bands at ~ 180 cm^{-1} and 260 cm^{-1} in the spectrum of poly (4-bromo-2,5-dimethyl-styrene) (Figure 1k) is beyond doubt, the use of the interval 150 – 350 cm^{-1} for qualitative analysis and when establishing various correlations should still be performed with a large with great care. A similar remark applies to the informativeness of the 16a vibration band, which appears near 405 cm^{-1} in the FIR spectra of PS and all its modifications presented here, regardless of the number, type, and position of the substituent in the benzene ring.

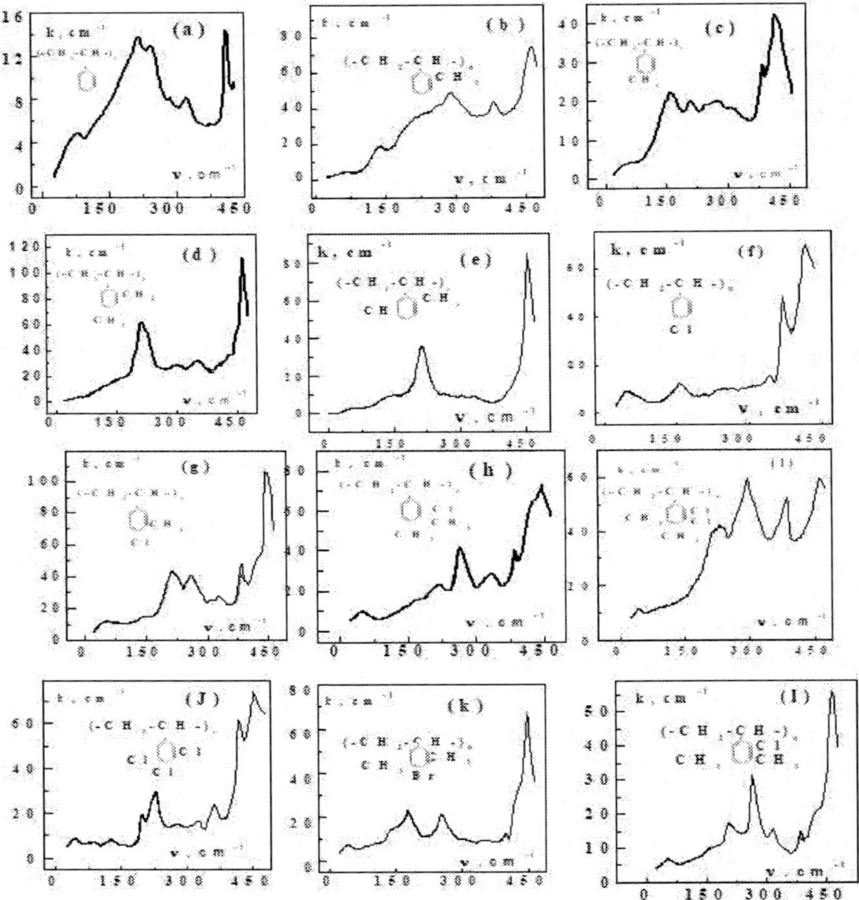

Figure 1. FIR spectra of polystyrene and its modifications. Poly-2,3-dichloro-4,5-dimethylstyrene (l); poly-2,4-dimethylstyrene (d); poly-2-methylstyrene (b); poly-2-chloro-3,5-dimethylstyrene (i); poly-2,4,5-trichlorostyrene (j); poly-4-bromo-2,5-dimethylstyrene (k); poly-4-chlorostyrene (f); polystyrene (a); poly-2-chloro-3,4-dimethyl-styrene (h); poly-4-methylstyrene (c); poly-2,5-dimethylstyrene (e), poly-4-chloro-3-methylstyrene (g).

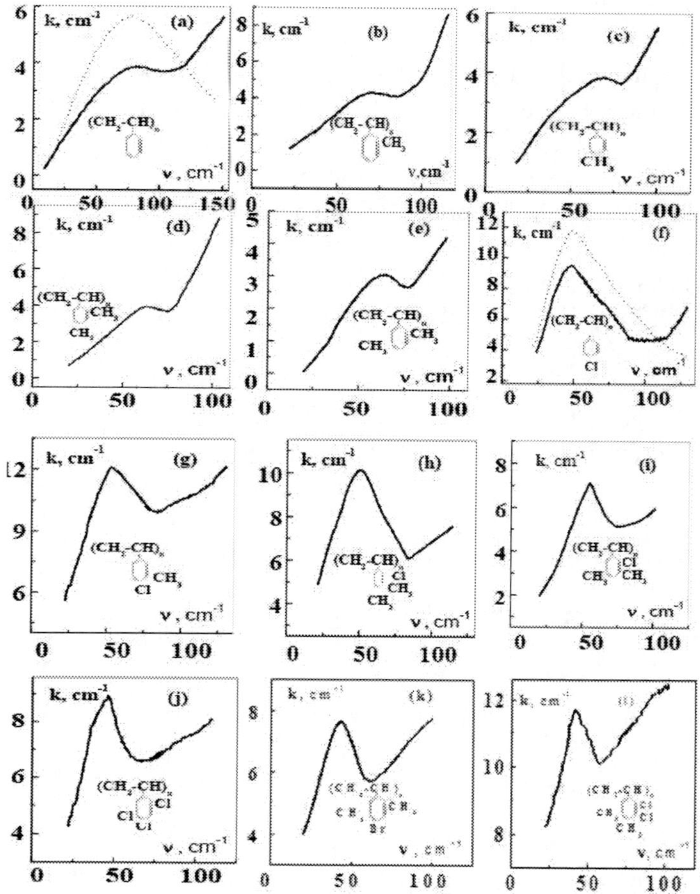

Figure 2. Poley-type absorption in the FIR spectra of polystyrene (a) and its modifications (b-l). The designations are the same as in Figure 1. Dotted lines in Figure 2a and 2f show the spectra of C_6H_6 and C_6H_5Cl, respectively.

In this regard, the use of spectral absorption parameters in the range below 150 cm^{-1} is more promising for analyzing and establishing the relationship between the parameters of the FIR spectra and the molecular characteristics of the studied polymers, since theoretical predictions exclude the participation of internal modes at these frequencies. It is now well established that the main component of the FIR spectra of polymers in the range below 150 cm^{-1} is Poley-type absorption [16]. This abnormally broad absorption in the frequency range from 10 to 100 cm^{-1} is not unique for polymers; it is also typical for liquids and many disordered solids, for example, for glasses [6, 17].

The appearance of absorption in the region below 150 cm_{-1} can be described by the model of damped rotational vibrations (librations) of polar molecules [5, 19–21]. One of the simplest models of molecular motion that is currently used to analyze this absorption is the model of limited rotators or libration of a molecule in the Bro - Darmon potential well [20]. According to this model, the libration motion of a polar molecule with the moment of inertia I occurs within the potential well formed by its closest environment, which has the form $U(\varphi) = V_0 Sin^2 \pi\varphi/2\xi$, where V_0 is the depth of the well, ξ is its semi-angular aperture (width at a height equal to half of the full barrier), and φ is the libration angle (amplitude of limited oscillations). The movement is performed with a circular frequency $\omega_0 = 2\pi s \nu_{libr} = 1/\pi\xi \cdot (V_0/2I)^{1/2}$. The integral absorption intensity is determined by the expression [21]: $A = \int k(\nu)d\nu = \pi (3c2)^{-1} \cdot \sum\mu_z^{\ 2} \cdot (1/I_x + 1/I_y)$, where $k(\nu)$ is the absorption coefficient, μ_z is the dipole moment along the molecular axis, I_x and I_y are the moments of inertia perpendicular to this axis.

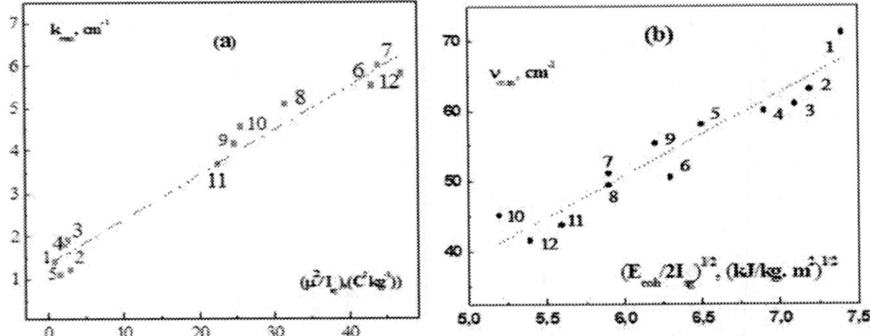

Figure 3. Dependence of the intensity of the “Poley-type absorption” band (a) and its frequency at the maximum (b) on the values of the molecular parameters of the polymers. The numbering of the points is the same as in Table 1.

The bands of this absorption in the FIR spectra of the studied PS derivatives are shown in Figure 2. As can be seen, it is an anomalously wide band (Δν = 30-50 cm-1), the intensity and position of which change significantly when the structure of the benzene ring is varied by substitutions. Depending on the type of substituent, their number and position in the benzene ring, the intensity of this absorption (kmax) can increase several times, and the position of its maximum (νmax) shift by 30-40 cm-1 compared to the FIR spectrum of the PS.

Table 2. Relationship between Poley-type absorption parameters and parameters of local segmental mobility in PS and its modifications. The polymer numbers correspond to those in Table 1

Polymer no.	Maximum Maximum Position ν_{max} (cm^{-1})	Intensity k_{max} (cm^{-1})	Barrier Q_{libr} – $=7.1 \cdot 10^{20} \cdot I_{бr} \cdot (\nu_{max})^2$ (10^{-19} J)	Cohesion energy E_{coh} (10^{-19} J)	$(E_{кor}/2I)^{1/2}$ (10^{10} (kJ/kg.m^2)$^{1/2}$)	$E_{кor}/Q_{либ.}$	Statistical segment S (number of monomers)	Tg, K (according to the DSC data of [13])	U_α, 10–19 J (according to the DSC data	$U_\beta = Q_{либ.} \cdot S$ (10^{-19} J)
1	72.0	1.4	0.152	0.60	7.4	3.0	8.0	372	1.64	1.3
2	63.0	1.2	0.160	0.73	7.2	3.5	8.0	396	1.74	1.3
3	61.0	1.9	0.150	0.71	7.1	3.7	8.0	386	1.70	1.2
4	60.0	1.8	0.175	0.80	6.9	3.7	8.0	409	1.80	1.4
5	58.0	1.1	0.170	0.72	6.5	3.4	8.0	392	1.73	1.4
6	50.5	5.5	0.130	0.73	6.3	4.3	11.0	405	1.76	1.4
7	51.0	6.0	0.150	0.72	5.9	3.8	8.0	394	1.73	1.2
8	49.4	5.1	0.177	0.83	5.9	3.8	9.0	410	1.80	1.6
9	55.3	4.2	0.220	0.91	6.2	3.5	8.0	420	1.85	1.8
10	45.1	4.6	0.187	0.84	5.2	3.6	9.0	410	1.80	1.7
11	43.8	3.7	0.184	1.00	5.6	4.5	10.0	426	1.87	1.8
12	41.5	5.6	0.166	0.97	5.4	4.8	11.0	422	1.86	1.8

The relationship between the kmax and νmax values with such molecular characteristics of the studied polymers as the effective dipole moment of the side group (μ), its geometry (moment of inertia I), and the cohesion energy E_{coh}, which characterizes the intermolecular interaction in polymers [22] are illustrated by graphs (a) and (b) in Figure 3.

Graph 3a shows that for the studied polymers there is a proportional relationship between k_{max} and μ^2/I, which directly indicates the librational nature of this absorption. In accordance with this assignment, the proportionality illustrated in graph 3b is found between ν_{max} and the value $(E_{coh}/2I)^{1/2}$, reflecting the dependence of the potential barrier for limited torsional vibrations (libration) on the level of intermolecular interactions in the studied polymers.

The revealed correlations make it possible, using the Bro - Darmon model, to estimate potential barriers to the libration of side groups in PS macromolecules and its modifications.

According to this approach, at small angles of torsional vibrations, the libration frequency (ν_{libr}) and the potential barrier for libration are related by the ratio: $\nu_{libre} = (c\pi\varphi)^{-1} \cdot (Q_{libre}/2I)^{1/2}$ [21].

Note that the Bro-Darmon model for polar liquids is in good agreement with experiment, and the potential libration barriers calculated using of this model are close to those obtained by other methods. [19.23].

Calculation of Q_{dibr} was carried out in two versions: under the assumption of the side group libration and when considering the monomer unit as a whole as a librator. At the calculation conventionally assumed the spherical van der Waals volume of the librator and the moment of inertia was determined by the formula $I = 0.4\ MR^2$, where M is the molecular weight and R is the radius side group. The libration angle φ was taken equal to 12.5^0 on the basis of theoretical models and calculations, which indicate the presence in macromolecules of limited torsional vibrations of atomic groups with an amplitude of 10 to 15° [24].

The results of calculating the libration barriers of side groups in PS macromolecules and its modification Q_{libr} are given in Table 2 (column 4), where they are compared with the cohesion energies Ecoh (column 5) of the side groups of the same polymers taken from [22]. Notable proximity of Q_{libr} to $E_{coh}/3$; it points to the intermolecular nature of the libration barrier. Note that the ratio $Q_{libr} \approx E_{coh}/3$ is also valid for simple liquids and that the value $E_{coh}/3$ corresponds to the intermolecular potential barrier for the displacement of molecular units in a condensed system relative to neighbors [25]. This type of mobility, first characterized by Reid and Evans [23] as a universal γ-process in liquid, is a precursor of the reorientation movement of molecules, i.e., β-relaxation. The β-process is a consequence of the influence of fluctuations in the nearest environment of a molecular unit on librational motion, which ensures its reorientation when passing from one potential well to another; for polymers, this relaxation process is an elementary event of the segmental (conformational) mobility of a macromolecule.

As can be seen from Table 2 (columns 4, 9, and 10), the potential barriers to libration in monomer units, calculated in this work using relation (1), are significantly (8-11 times) lower than the barriers for elementary acts of segmental motion (U_a), calculated in [10], proceeding from the glass transition temperatures (T_g)) of the PS and its modifications.

In [10], changes in glass transition temperatures (and, consequently, changes in barriers to segmental motion) in the case of modifications of the benzene ring with substituents were associated with changes in the mass M (steric conditions of mobility) and with changes in intra- and intermolecular interactions, which were characterized by UV spectra and dipole moments (μ) of modified benzene rings. Taking into account the relationship of these factors (steric and "electronic") made it possible to explain why, with the same mass (M), in the case of samples 8 and 9 (Tables and 2), these polymers have different glass transition temperatures and barriers to segmental motion: their monomeric units have different dipole moments (μ). At the same time, samples 8 and 10 with different molecular mass of the benzene ring, have the same glass transition temperature, since they have similar dipole moments of monomer units (Table 1, column 7).

However, in our opinion, it is necessary to take into account the polymer specificity of segmental mobility, which consists in the correlation of the motion of neighboring dipoles at fixed bond angles and inhibition of internal rotation in the units of the macromolecule.

In other words, the local conditions of segmental motion (and its potential barrier) are determined not only by the parameters of monomeric units (size, mass, dipole moment, optical anisotropy, etc.), but also by the average mutual arrangement of the latter, which depends on the flexibility of the macromolecule [26, 27]. We believe that if in low-molecular-weight liquids the mobility responsible for β-relaxation is the rotational-translational displacement of molecules, then in polymers, this is the reorientation movement of a section of the chain, which is statistically independent of the orientation of adjacent regions of the chain. Such a statistical segment (or Kuhn's segment), which characterizes the thermodynamic flexibility of the polymer chain, in the case of PS includes 7-8 monomer units [9], and for substituted PS, according to calculations [28], it is equal to 7-11 monomeric units. The values of the statistical segments (S) of the substituted PSs studied in this work, found in the literature [22, 29] or calculated by the method [28], are given in Table 2, column 8.

Knowing the size of the statistical segment (*S)* of the studied polymers, we can estimate the height of the barrier for local segmental motion using the dependence of the form: $U_\beta = (0.3 \pm 0.05)\, E_{coh} \cdot S + B$ obtained in [9] for the β-process. With a relatively small value of the term $B \approx 10 \pm 5$ kJ/mol, corresponding to the barrier of internal rotation around the C - C bond in flexible-chain polymers and the approximate equality $E_{coh} \approx 3Q_{libr}$, we have a formula by which these barriers were calculated in the present study: $U_\beta = Q_{libr} \cdot S$. As can be seen from Table 2 (columns 10 and 11), the activation energies of β-relaxation in PS and its modifications calculated on the basis of the obtained data are comparable with the barriers for elementary acts of segmental motion in the same substituted polystyrenes, estimated in [10] by their glass transition temperatures.

The results obtained confirm the mechanism proposed in [9] for β-relaxation, which consists in the rotational movement of the links of the chain section, which is close in magnitude to the correlation section, the statistical segment, with overcoming predominantly intermolecular barriers and with the participation of a single-barrier trans-gosh transition. This act of motion is similar to the act of segmental relaxation in a polymer melt and can be realized in a solid polymer as an elementary act of cooperative conformational rearrangements in the α-process.

The closeness of the barriers of correlated librational movement (U_β) and the barriers of local segmental mobility (U_α) determined by T_g by DSC, revealed in this work, confirms that the activation barrier of conformational mobility in the polymer chain changes little in a wide temperature range $T > T_\beta$, only due to changes in interchain interactions, while remaining close to Q_β.[3]

Conclusion

In this work, we measured and analyzed the FIR spectra of PS and its modifications obtained by substituting methyl groups and/or chlorine and

---

[3] With the degeneration of the cooperative nature of mobility at high frequencies ($\upsilon \geq 10^7$ to 10^8 Hz), both processes are combined into a single $\alpha\beta$-process with activation energy U_β.

bromine atoms for hydrogen atoms in the benzene rings of the side groups of macromolecules.

All these spectra have an anomalously wide asymmetric absorption band with a maximum position in the range from 40 to 80 cm^{-1}, depending on the nature of the substituent. This absorption band can be attributed to absorption associated with libration (rotational vibrations) of phenolic rings, that is, to absorption by the Poley mechanism.

The relationship between the spectral absorption parameters of Poley and the molecular characteristics of the studied polymers allows direct to analyze the role of molecular structure and intermolecular forces in their dynamics. Height of potential barriers to libration of phenolic rings in PS macromolecules and its modifications Q_{libr}, estimated on the basis of the analysis of this absorption, were found to be close to the activation barriers of low-temperature mechanical β-relaxation in these polymers. Comparison of these barriers with the activation barriers of local segmental movement, determined by the DSC method for the same polystyrenes, showed that in all cases the equality $U_\beta \approx Q_{libr} \cdot S$ (S-length of the statistical segment) takes place, confirming that the β-process, as an elementary act segmental movement in polymers is associated with correlative librational movement of a chain section, which is statistically independent. In this sense, the universal δ-process, characterized here by the absorption by the Poley mechanism in the FIR spectra, is a high-frequency precursor of the β-process.

In general, the results presented above show that low-frequency infrared spectroscopy makes it possible to specify the molecular mechanisms of secondary relaxations in polymers and to establish the relationship of these processes with the molecular characteristics of polymers, such as the structure of the monomer unit, cohesion energy, potential barrier to internal rotation, and thermodynamic chain stiffness.

References

[1] Gottlieb Yu.Ya., Darinsky A.A., Svetlov Yu.E. *Physical kinetics of macromolecules.* L., 1986.

[2] Johari G.P.//*J. Non-Cryst. Solids.* 2002. V. 307-310. P.114.

[3] Coffey W., Evans M., Grigolini P. *Molecular diffusion and spectra.* M., 1987.

[4] Ding Y., Kisliuk A., and Sokolov A.P.//*Macromolecules.* 2004. V.37. P.161.

[5] Bershtein V.A., Ryzhov V.A.//*Adv.Polym.Sci.* 1994. V.114. P.43.

[6] Slutsker A.I., Vasilyeva K.V., Egorov V.M., Dokukina A.F.//*Vysokomol. connect.* 2002.vol. 44. p. 2103.

[7] Dokukina A.F., Coton M.M.//*Vysokomol. connect.* 1959.Vol. 1, p. 1129.
[8] Dokukina A.F. ..., Coton M.M.//*J. of general chemistry.* 1959.vol. 29. p. 2201.
[9] Varsanyi G. *Vibrational spectra of benzene derivatives.* New York. 1969.
[10] Spells S., Shepherd I. W.//*Polymer.* 1977. V.18. P. 906.
[11] Gribov L.A. *Theory of infrared spectra of polymers.* M. 1972.
[12] Jasse B and Monnerie L.//*J. Phys. D: Appl. Phys.* 1975. V.8. P. 863.
[13] Hill N., Vangan W., Prise A., Davies M. *Dielectric properties and molecular behavior.* Van Nostrand, Princeton, New York. 1969.
[14] Van Krevelen D.W. *Properties of polymers. Correlation with chemical structure.* Elsevier, Amsterdam. 1972.
[15] 1Askadsky A.A., Matveev Yu.I. Chemical structure and physical properties of polymers. *M. Chemistry.* 1983.
[16] Polymer Handbook/Ed. by J. Brandrup, E. Immergut. New York. 1975.
[17] Darmon I., Brot G.//*Molec. Cryst.* 1967. V.2 P. 36.
[18] Birshtein T.M., Ptitsin O.B. *Conformation of macromolecules.* M. 1984.
[19] Glasston S., Leidler K., Eyring G. *Theory of absolute velocities reactions.* M. 1948.
[20] Bershtein V.A., Egorov V.M. *Differential Scanning Calorimetry of Polymers.* New York. 1994.

2. Low-Frequency Librational Oscillations, "Boson Peak" and Interchain Interactions in a Glassy Polymer

The IR and Raman spectra of plasticized polymethyl methacrylate in the range of 10-120 cm^{-1} were obtained and analyzed, where absorption and scattering, caused by both individual (85-90 cm^{-1}) and correlated librational vibrations (15-20 cm^{-1}, "Boson peak") preceding the manifestation of relaxation dynamics. It is shown that the presence of a "boson peak" in low-frequency spectra as a sign of the solid-state behavior of a polymer substance is due to the correlation of librational vibrations not only within the macromolecule (in a region comparable with Kuhn's segment), but also in segments of adjacent chains.

Introduction

Low-frequency vibrational excitations in flexible-chain polymers and other disordered systems are actively emitted using Raman, IR, and neutron spectroscopy [1-4]. Special attention of researchers is attracted by the characteristic absorption band in the region of 10-50 cm^{-1}, caused by such excitations and called the "boson peak" (BP) in Raman spectra.

The manifestation of BP in the spectra of liquids during their cooling is considered the first sign of solid-state behavior of a disordered system. The BP is fixed even before the complete glass transition of the liquid at the characteristic temperature Tc in the coupled mode model [5] ($T_c > T_g$, where T_g is the glass transition temperature), which for polymers is close in value to the liquid-liquid transition temperature $T_{ll} \sim 1.2T_g$ (according to Boyer [6]). Essentially, T_c in linear polymers corresponds to the transition temperature from the viscous to the highly elastic state.

To explain the nature of BP, a number of models have been proposed that associate it with structural defects [7], fractals [8], or localized excitations in the model of soft potentials [9]. There are also models in which the existence of a cohesive inhomogeneity in medium-order disordered media is assumed to be responsible for the manifestation of the BP. The idea of Shuker and Gamon [10] that low-frequency vibrational excitations are localized at nanoscale inhomogeneities of the structure was developed by Martin and Brenig [11], Duval [12] and Malinovsky [13]. According to [11-13], the dynamic correlation size L of the oscillating unit responsible for the BP in the Raman spectrum is

$$L \approx c_t\,(c \cdot \nu_{BP})^{-1} \approx -5\text{nm},$$

where c is the speed of light in vacuum, c_t is the "transverse" speed of sound (shear wave velocity), ν_{BP} is the frequency of the BP maximum.

At present, the existence of a connection between correlated vibrational excitations and relaxation segmental dynamics in polymers seems to be undeniable [2, 14-17]. In our previous work [18], we performed experiments by the methods of far IR (FIR) and Raman spectroscopy on samples of polymethyl methacrylate (PMMA) with different predetermined molecular lengths (mainly oligomers). It was shown that the position of the BP in the spectra depends on the chain length and the linear size of the region of coherence of librational excitations in the polymer is comparable with the value of the statistical Kuhn segment of the macromolecule. This result revealed the important role of the intramolecular correlation of low-angle librational excitations in the polymer and indicated their genetic relationship with large-amplitude (conformational) segmental dynamics.

Indeed, it was experimentally established [19, 20] that at all temperatures, starting from the β- transition temperature $T_\beta < T_g$, the relaxation dynamics in flexible-chain polymers is realized with the participation of conformational transitions [21] and is controlled by acts of kinetically independent (*β* - process) or intermolecularly cooperative (*α* - process near T_g) segmental movement; the unit of motion in this case is a chain section commensurate with the Kuhn segment of the polymer. This segment with a length of 2-3 nm typical for flexible-chain polymers characterizes in this case the dynamic correlation of neighboring units in the macromolecule. Relaxation dynamics in the glass transition interval and at temperatures close to Tg is realized with the involvement of 3 - 5 adjacent segments of adjacent chains in the acts of motion (at low frequencies) [19]. Therefore, the transverse size of the cooperatively tunable region, as a rule, is also commensurate with the length of the statistical segment in flexible-chain polymers.

At the same time, the intramolecular dynamic correlation persists even at temperatures above T_c, ie, in a polymer melt [19], but BP appears in the spectra only at temperatures $T < T_c$. This suggests that, along with the intrachain correlation of librational excitations, an important interchain interaction can also play a role in the formation of the low-frequency vibrational spectrum in the range 10-50 cm^{-1}. In other words, it is natural to assume not a one-dimensional, but a three-dimensional nature of the nanoregion of correlated excitations that determine the manifestation of BP.

The aim of this work was to experimentally verify this assumption by FIR and Raman spectroscopy on amorphous PMMA, plasticized in various ways with dibutyl phthalate (DBP). Molecules of the latter, "pushing" the chains, reduce and even completely eliminate interchain interactions in PMMA. In this case, DBP has approximately the same solubility parameter δ ($\delta = (E/V)^{1/2}$, where E is the cohesion energy, V is the molar volume, and E/V is the cohesion energy density) as PMMA. As a result, plasticization leads to a decrease in the degree of interchain cooperativity of the movement of polymer segments (up to the loss of cooperativity), but the quasi-independent movement of segments will occur in the same cohesive field as in the absence of a plasticizer.

Materials and Methods

Film samples of plasticized PMMA were prepared by casting onto glass a solution of polymer and DBP in a toluene-dioxane-dichloroethane-acetone mixture. We used PMMA with a number-average molecular weight $Mn \approx$

500000 g/mol. Investigated unplasticized PMMA and its compositions with 9, 17 and 23 wt.% DBP. Their glass transition temperatures, measured by differential scanning calorimetry at a heating rate of 20 K/min, were 380, 345, 328, and 310 K, respectively. spectra samples of the same compositions made in the form of parallelepipeds with polished edges measuring 40 x 10 x 10 mm. All measurements were carried out at 293 K.

FIR spectra were recorded on two spectrometers: in the range from 10 to 50 cm^{-1} on an LGU spectrometer [22] and in the range 550 - 120 cm^{-1} using the FIS-21 Hitachi device. The resolution was - 2 cm^{-1}. The accuracy of determining the position of the band maxima was 2 - 3 cm^{-1}. The error in measuring the absorption coefficient $k = ln\ (T^{-1})/d$ (where T is the transmission of a sample of thickness d) was 5-10%. The absorption by the plasticizer itself did not distort the FIR spectra, since it was negligible.

Raman spectra were recorded on a double Ramalog-5 monochromator using a 90° scattering scheme and were obtained using an argon laser with a power in the 488 nm line of about 0.1 W. The spectral slit width was 2 cm^{-1}. The scattering intensity spectra $I\ (v)$, averaged over ten scans in the range 5-120 cm^{-1}, were normalized taking into account the spectral function of the device. The calculation of low-frequency Raman spectra, based on the obtained experimental spectra, was carried out by transforming the latter according to the formula widely used in the literature [23-25]: $I_R\ (v) = I\ (v)\ [1 - exp\ (-hv/k_BT)] = I\ (v)\ [N\ (v) + 1]^{-1}$, where $N(v)=[exp\ (hv/k_BT) - 1]^{-1}$ is the Bose factor.

The representation of the Raman spectrum in coordinates $I\ (v)/[N\ (v) + 1]$ has the advantage that it allows one to significantly reduce the contributions of the elastic (Rayleigh) and relaxation components of the scattering wing. In addition, the recalculated spectra are directly comparable with the spectra of dielectric losses $\varepsilon''(v)$. Since at high frequencies the absorption coefficient k is proportional to $2nv\varepsilon''(v)$, the spectrum $I\ (v)/[N\ (v) +1]$ also corresponds to the FIR spectrum in the coordinates $k\ (v)/v$. Recalculated in this way, the Raman and FIR spectra can be correctly correlated with each other.

Results and its Discussion

Experimental Raman and FIR spectra of plasticized and unplasticized PMMA samples are shown in Figure 1, and their reduced spectra in coordinates $I\ (v)/[N\ (v) + 1]$ and $k\ (v)/v$ are shown in Figure 2.

Figure shows that the experimental Raman and FIR spectra of the initial PMMA differ in many respects. What they have in common is the manifestation of a wide band with a maximum at ~ 90 cm^{-1}. In the Raman spectrum of PMMA, one can also see the BP at 15–20 cm^{-1} and a sharp increase in scattering at frequencies below 10 cm^{-1}, caused by the manifestation of the relaxation component of the spectrum (β relaxation [26]).

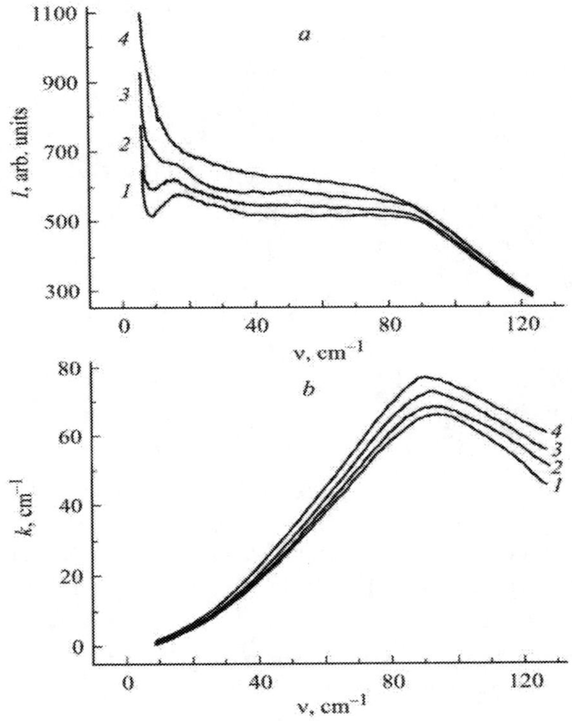

Figure 1. Experimental Raman (a) and FIR spectra (b) in the range 10 - 120 cm^{-1} of the initial PMMA (1) and PMMA plasticized with DBP (2-4) at 293 K. PMMA/DBP, wt .%: 2 - 91/9, 3 - 83/17, 4 - 77/23.

With plasticization in the Raman spectrum of PMMA, it is the relaxation component that increases at frequencies in the BP region and below. An increase in the DBP concentration leads to the fact that the minimum characterizing the transition from the relaxation component of the spectrum to the vibrational (BP) one gradually disappears on the spectral curve. As a result, at 23 wt.% DBP in the sample, BP ceases to appear.

In the experimental FIR spectra, the BP is not represented by a separate band. However, a comparison of the presented Raman and DIR spectra of the studied samples (Figure 2) indicates a close agreement of such spectra in the range of 10 – 120 cm^{-1}. In position and shape, they have a wide asymmetric band with a maximum at 80–90 cm^{-1} and a bend (kink) at the frequency of the boson peak at 15–20 cm^{-1}, typical of the low-frequency spectra of linear amorphous polymers.

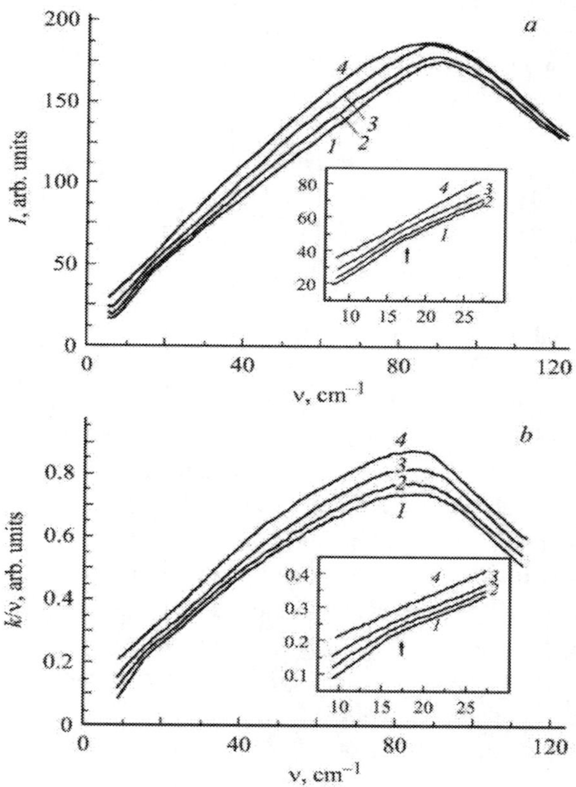

Figure 2. Presented Raman- (a) and FIR-spectra (b) in the range of 10-120 cm^{-1} of the initial (1) and plasticized (2-4) PMMA. The notation of the curves is the same as in Figure 1. The insets show the initial portions of the curves on an enlarged scale, the arrows show the position of the boson peak in the spectra.

Figure 2 shows that, as plasticization progresses, the main band in the low-frequency spectrum of PMMA broadens somewhat and increases in intensity and rises at 5-10 cm^{-1} to low frequencies. The latter is obviously due to an increase in the amplitude and anharmonicity of oscillations due to the weakening of interchain interactions. Of particular interest is the practical disappearance of the kink in the BP region (15 - 20 cm^{-1}) for plasticized PMMA with 23 wt% DBP in shown Raman and FIR spectra (Figure 2).

It was previously found that the main band in the FIR spectra at frequencies below 100 cm^{-1} refers to absorption by the Poley mechanism, ie, to small-angle librational motions (rotational oscillations) of molecules in the case of low-molecular substances [27, 28] or monomeric units in macromolecules [20, 29, 30]. The same motion also includes the band at 80 - 90 cm^{-1} in the Raman spectrum of PMMA, which has a similar profile and a position of the maximum close in frequency [31].

On the other hand, as noted above, the position of the inflection on the low-frequency wing of the libration band in the Raman and FIR spectra of PMMA is determined by the size of the librator. Experiments [18] showed that this inflection is at the BP frequency at 15 – 20 cm^{-1} in the FIR spectra of PMMA only if the chain length exceeds 2 nm. In molecules of shorter length, the BP shifts to higher frequencies: for example, on going from the seven-dimensional to the dimer, this shift was ~ 30 cm^{-1}. The found dependence of the position of the inflection of the spectral curve on the chain length made it possible to relate absorption and scattering at the BP frequency with correlated librations of units in a portion of the molecular chain comparable in length with a statistical segment. The latter characterizes in this case the dynamic interrelation of neighboring units in the chain and is equal for PMMA to six monomeric units [19].

The absence of an inflection at the BP frequency, observed in the presented Raman and FIR spectra upon the introduction of 23 wt% DBP into PMMA (Figure 2), provides, in our opinion, additional information for understanding the nature of BP in polymers, namely, confirms the role of interchain interactions in the manifestation of BP.

It should be noted that the observed effect refers to a system with a liquid plasticizer (for DBP, Tg = 190 K), in contrast to previous studies on the PMMA-DBP system performed by Raman spectroscopy at 140 K [32] and neutron spectroscopy at 30 K [33], ie, under conditions of suppression of segmental relaxation dynamics. Under these conditions, the solid-state BP was preserved and became even more pronounced. Indeed, the collective (correlated vibrational excitations are retained in the polymer system up to helium temperatures [12, 23]. With increasing temperature, their amplitude increases, and under certain conditions, they become the direct predecessors of the emergence of acts of segmental, relaxation dynamics.

In this respect, it is interesting to compare changes in the low-frequency spectrum of librational excitations in PMMA with changes in relaxation dynamics caused by an increase in temperature or plasticization.

With the help of differential scanning calorimetry and other methods, the general nature of the glass transition (α-transition) and β-relaxation ($T_\beta < T_g$) as, respectively, intermolecular-cooperative and quasi-independent movement of sections of polymer chains, commensurate with statistical segment [19, 20]. Near T_g, the parameter of interchain cooperativity of segmental movement at low frequencies is $Z \approx Q_\alpha/Q_\beta \sim V_\alpha/V_\beta \approx 4 \pm 1$, where Q_α *and* Q_β are the effective activation energies of these processes, and V_α and V_β are their activation volumes.

Heating the polymer above T_g leads to a gradual decrease in the cooperativity of segmental movement due to weakening of intermolecular contacts and an increase in free volume; starting from the temperature T_c, the relaxation dynamics is realized as a quasi-independent motion of the segments [19]. A similar "uncooperation" with a decrease in the scale of the act of motion at T_c by a factor of 3–4 is also predicted in the model of coupled modes of the theory of glass transition [5, 34].

This effect can be achieved in another way - plasticization of polymers [19, 35]. In the limit, plasticization, which violates interchain interactions, reduces T_g down to the temperature T_β, the value of Q_α - to the activation energy of the β-transition Q_β, the activation volume V_α - to the volume of the statistical segment V_β. It was shown [35] that in the PMMA-DBP system, the cooperative motion of polymer segments in the glass transition is replaced by their quasi-independent motion ($Z \approx 1$) even at a DBP content of ≈ 23 wt% (Figure 3a).

To compare the changes observed in the low-frequency PMMA spectra with increasing temperature or due to plasticization, we use the criterion of Sokolov et al. [24], which makes it possible to estimate the relaxation contribution to the Raman spectrum from the ratio of the intensity at the BP frequency to the intensity in minimum (≈ 10 cm^{-1}) of the spectral curve: $R = I_{BP}/I_{min}$.

This ratio turned out to be associated, in particular, with such a characteristic as the rate of decrease in the average relaxation time (τ) with increasing temperature - the parameter throm $m = -d(\log(\tau))/d(T_g/T)$. The larger the parameter m, the smaller the value of R and the more significant the relaxation contribution to the spectrum. In fact, with temperature, the relaxation contribution to the spectrum grows faster than the vibrational one, and at T_c it completely overlaps the BP, indicating the final transition of the system to a liquid-like state.

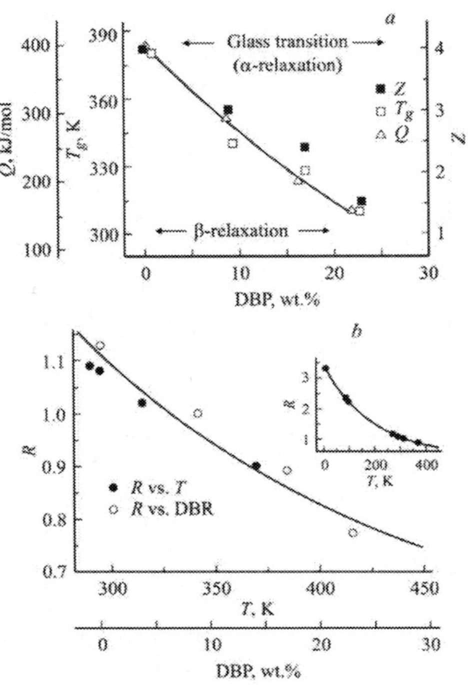

Figure 3. (a) Dependences of the parameters *Tg*, *Q*, and *Z* for plasticized PMMA on the concentration of DBP and (b) the dependence of the parameter *R* on temperature (for the initial PMMA) (data [23, 36]) and on the concentration of DBP in PMMA.

In Figure 3b shows a graph of the variation of the parameter *R* with temperature for PMMA, constructed from the data of [23, 36]. In the temperature range from 293 K to T_g + (70 - 100°), the experimental points were satisfactorily described by the temperature dependence of the parameter m with the values characteristic of PMMA in the expression for the relaxation time $\tau = \tau_0 \exp (BT_0)/(T - T_0)$: $m_0 \approx 10^{-13}$ s, $T_0 \approx 285$ K, $T_g \approx 380$ K and coefficient $B \approx 4$ [37]. The same figure shows the dependence of the parameter *R*, found from the low-frequency spectra of plasticized PMMA, as a function of the plasticizer content. It can be seen that the experimental points actually fall on the graph of the variation of the parameter *R* with temperature. In this case, the value of *R* obtained from the spectra of a sample with 23 wt% DBP at 293 K corresponds to the ratio $R = I_{BP}/I_{min}$ in the spectra of unplasticized PMMA at a temperature of - 425 K, which is close to the temperatures T_{ll} (according to Boyer) and T_c in the model of coupled mode equal to - 435K for PMMA [38].

Conclusion

Thus, this work shows the equivalence of the effect of plasticization and temperature increase on the Raman and FIR spectra of an amorphous polymer in the range 10 - 120 cm^{-1} vibrational (librational) type of motion at BP frequencies to the relaxation dynamics of the β-process.

An analysis of the FIR and Raman spectra obtained in the present and previous works [17, 18] confirms the idea that BP in polymers is determined by acts of correlated librational excitations in nanoregions with increased cohesion at intramolecular) and interchain cooperation. The latter is confirmed by the disappearance of the characteristic inflection at 15 - 20 cm^{-1}, which corresponds to the BP, in the presented FIR and Raman spectra of plasticized PMMA under the conditions of the decomposition of the blocks of segments and the loss of the solid state by the flexible-chain polymer.

References

[1] Malinovsky, V.K., V.N. Novikov, A.P. Sokolov. *UFN* 163, 119 (1993).
[2] Johari, G.P. *J. Non-Cryst. Solids* 307-310,114 (2002).
[3] Angell, C.A. J. Phys .: Cond. Matter 16, 85 153 (2004).
[4] Wypych, A., E. Duval, L. David, A. Mermet. *Polymer* 46,12 523 (2005).
[5] Gotze, W., L. Sjogren. *Rep. Prog. Phys.* 55, 241 (1992).
[6] Boyer, R.F., R.L. Miller. *Macromolecules* 17, 365 (1984).
[7] Elliot, S.R. *Europhys. Lett.* 19, 201 (1992).
[8] Bagryansky, V.A., V.K. Malinovsky, V.N. Novikov, L.M. Pushchaev. *FTT* 30, 2360 (1988).
[9] Buchenau, U., Y.M. Galperin, V.I. Gurevich, D.A. Parshin, M.A. Ramos, H.R. Schrober. *Phys. Rev.* B 46, 2798 (1992).
[10] Shuker, R., R.W. Gammon. *Phys. Rev. Lett.* 25, 122 (1970).
[11] Martin, A.J., W. Brenig. *Phys. Status Solidi* 64, 163 (1974).
[12] Mermet, A., N.V. Surovtsev, E. Duval, J. Jal, J. Dupuy-Philos, A. Dianous. *Europhys. Lett.* 36, 277 (1996).
[13] Malinovsky, V.K. *FTT* 41, 805 (1999).
[14] Novikov, V.N. *J. Non-Cryst. Solids* 235-237,196 (1998).

[15] Schober, H.R. *J. Phys .: Cond. Matter* 16, 52 659 (2004).

[16] Ngai, K.L. *J. Non-Cryst. Solids* 353, 4237 (2007) 31, 475 (1993).

[17] Bershtein, V.A., V.A. Ryzhov, L.M. Egorova, E.V. Kober. 3rd Int. *Meeting on relaxation in complex systems.* Vigo, Spain (1997). P. VI-4.

[18] Ryzhov, V.A. *FTT* 44, 2229 (2002).

[19] Bershtein, V.A., V.M. Egorov. Differential scanning calorimetry in physicochemistry of polymers. Khimiya, L. (1990). 255 p. V.A. Bershtein, V.M. Egorov. *Differential scanning calorimetry of polymers. [Physics, chemistry, analysis, technology.* Ellis Horwood, N.Y. (1994). 253 p.].

[20] Bershtein, V.A., V.A. Ryzhov. *Adv. Pol. Sci.* 114, 43 (1994).

[21] Ryzhov, V.A., V.A. Berstein. *High molecular weight. compounds* A 31, 2482 (1989).

[22] Ryzhov, V.A., M.V. Tonkov. *In collection: Molecular Spectroscopy/*Ed. M.O. ulanin. Publishing house of Leningrad State University L. (1973). P. 103.

[23] Novikov, V.N., A.P. Sokolov, B. Stube, N.V. Surovtsev, E. Duval, A. Mermet. *J. Chem. Phys.* 107, 1057 (1997).

[24] Sokolov, A.P., A. Kisliuk, D. Quitmann, A. Kudlik, E. Rossler. J. Non-Cryst. Solids 172-174, 139 (1994).

[25] Caliskan, G., A. Kisliuk, V. Novikov, A. Sokolov. *J. Chem. Phys.* 114,10189 (2001).

[26] Malinovsky, V.K., V.N. Novikov, A.P. Sokolov. *FHS* 22, 204 (1996).

[27] Poley, J.P. *J. Appl. Sci.* B 4, 337 (1955).

[28] Evans, M.W., G.J. Evans, W.T. Coffey, P. Gricolini. *Molecular dynamics theory of band spectra.* Wiley-Interscience, N. Y. (1982). 866 p.

[29] Chantry, G.W., J.W. Fleming, E.A. Nicol, N.A. Willis, M.E.A. Cudby. *Infrared Phys.* 12, 101 (1972).

[30] Ryzhov, V.A., V.A. Berstein. High molecular weight. compounds A 31, 451 (1989).

[31] Viras, F., T.A. King. *Polymer* 26,899 (1984).

[32] Duval, E., M. Kozanecki, L. Saviot, L. David, S. Etienne, V.A. Berstein, V.A. Ryzhov. *Europhys. Lett.* 44, 747 (1998).

[33] Saviot, L., E. Duval, J.F. Jal, A.J. Dianoux, V.A. Bershtein, L. David, S. Etienne. *Phil. Mag.* B 82, 533 (2002).

[34] Novikov, V.N., A.P. Sokolov. *Phys. Rev.* E 67,031 507 (2003).

[35] Bershtein, V.A., L.M. Egorova, V.M. Egorov, A.B. Sinani. *High molecular weight. compounds* A 31, 2482 (1989).

[36] Surovtsev, N.V., T. Achibat, E. Duval, A. Mermet, V.N. Novikov. *J. Phys .: Cond. Matter* 7, 8077 (1995).

[37] Bohmer, R., K.L. Ngai, C.A. Angell, D.J. Plazek. *J. Chem. Phys.* 99, 4201 (1993).

[38] Murthy, S.N. *J. Polym. Sci. B: Polum. Phys.* 31, 475 (1993).

3. Technique of Far Infrared and Terahertz IR Spectroscopy

Far infrared region of the optical spectrum between the ordinary infrared region and the microwave radio range for a long time remained inaccessible for conducting spectral studies in this wavelength range. This is how the area from 25 μm to 1000 μm was allocated, (i.e., from 400 cm^{-1} to 10 cm^{-1}) into a special spectral region, different from the higher-frequency boundary region, which has long been used in physical chemistry to solve many different tasks. The first long-wavelength IR absorption spectra of polymers were obtained only in the middle of the last century, when high-aperture diffraction gratings appeared and highly sensitive radiation detectors and low-noise amplifying devices capable of receiving weak signals were created.

For research in the far infrared region, a large number of fairly good spectrometers with diffraction gratings have been designed [1-3].

The main experimental difficulties faced by the researcher when work in the spectral range from 400 to 10 cm^{-1}, consisted in the following. Sources of long-wave radiation used in the study of absorption spectra emit most of the energy in the visible and mid-infrared regions. This fact complicates the suppression of high grating orders and complicates the solution of an already complex problem - filtering radiation in the long-wavelength region. The strong absorption of radiation by atmospheric water vapor forces the researchers to evacuate the device, or at least fill it with dry gas. The relatively large width of this area creates an additional difficulty in work. In order to obtain a spectrum of a sample from 200 to 10 cm^{-1}, at least five different combinations of filters and gratings must be used.

In contrast to the radiation filtration system, the structures of the dispersing part of long-wave monochromators do not have any specific features in comparison with devices in the middle and near-infrared region; their only characteristic feature is their large aperture ratio.

Monochromators used in spectrometers for long-wavelength region are assembled according to the same schemes as monochromators for devices in the mid-IR region. These are off-axis and axial autocollimation schemes [4, 5], the Pfund scheme [6] and Ebert-Fasty [7]. More often than others, the Czerny - Turner scheme is used [8], which we will dwell on in more detail. This scheme is one of the modifications of the Ebert - Fasty scheme and differs from the latter in that instead of one spherical mirror, it uses two: one for collimation and one for focusing radiation onto the exit slit. This separation gives the scheme additional degrees of freedom in the fight against aberrations, while retaining all the advantages inherent in the Ebert-Fasty scheme. Full compensation of the coma in the Ebert-Fasty scheme is impossible due to the meridional increase in the lattice. Therefore, in practice, they often use the scheme Czerny - Turner with asymmetrical ray path. For registration of radiation in most spectral instruments in the long-wave region, pneumatic receivers are used, the long-wave region does not introduce any specificity into their work. A number of models used radiation detectors cooled to helium temperature. The optical layout of a typical spectrometer for operation in the far IR range is shown in Figure [9].

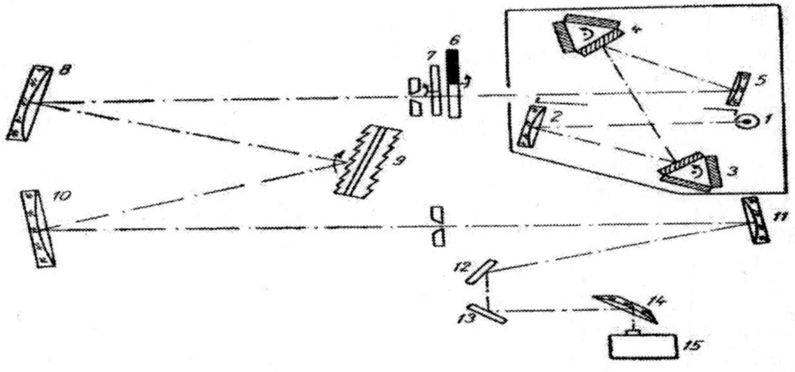

Figure 1. Optical layout of the device. - radiation source; 2, 5, 8, 10, 11 - spherical mirrors; 3, 4, 7 –filters, 6 - modulator; 9 - echelette; 12. 13 - flat mirrors; 14 - elliptical mirror; 15 - radiation receiver.

About a quarter of the FIR spectra of polymers presented in this review were obtained using such and similar spectrometers (for example, the Hitachi FIS-21 spectrometer or the Jasco DS-501 spectrometer).

A big step in mastering the far infrared range was the development of Fourier spectrometers that effectively use the wide-range radiation of heat sources [10]. Scanning two-beam interferometers serve as a dispersive element in Fourier spectrometers. The particular success of Fourier spectroscopy for the FIR range is largely due to the fact that, using systems of reasonable complexity and dimensions, it is possible to scan the interferometer mirror at a distance L much larger than the wavelength of the working radiation with a position reading error $\Delta L \ll \lambda$. This provides a high resolution Fourier transform spectrometer. In principle, it can reach 10^{-2} - 10^{-3} cm^{-1}, but the recording time for FIR spectra with such a resolution takes tens of hours.

In contrast to dispersive spectrometers, the principle of constructing Fourier spectroscopy instruments is based on the use of interference technology. An interference curve or an interferogram $I\ (\Delta)$ is recorded in it using a two-beam interferometer, described in the case of radiation of arbitrary composition $S\ (\nu)$ by the relation $I\ (\Delta) = \int S\ (\nu)\ cos2\pi\nu\Delta d\nu$, where Δ is the path difference of the interfering beams. Then, applying the integral Fourier transform to the function $I\ (\Delta)$, the desired spectrum $S\ (\nu) = \int I\ (\Delta)\ \cos 2\pi\nu\Delta d\Delta$ is obtained, expressed in terms of the measured value $I\ (\Delta)$. In general, the interferogram is measured as a function of the path difference and the spectrum is calculated using the relation $S\ (\nu)$ by an analog or digital method. The main difficulties of spectroscopy using the Fourier transform are associated precisely with the calculation of the spectrum. Therefore, the widespread use of this method was determined by the development of computer technology and a computer is a necessary element of a modern Fourier spectrometer.

As noted above, the main difficulty when working at low frequencies is associated with with low power heat sources in the FIR region. The intensity of blackbody radiation in the long-wavelength region of interest to us decreases as ν4, as a result, the total power of the heat source ($T = 400^0K$) is only a few microwatts for the entire FIR range ($\nu <100\ cm^{-1}$). To obtain an acceptable measurement accuracy under such conditions, it is necessary to use highly sensitive germanium bolometers ($S = 5 \cdot 10^{-13}\ W/Hz^{1/2}$) as radiation receivers, operating at a temperature of $T = 1.2$ K.

The advantages of Fourier spectroscopy over other spectroscopic methods using decomposition into a spectrum are determined, first of all, by the energy gains. The first is that the entrance aperture of Fourier spectrometers is much larger than that of dispersive devices, into which light enters through a narrow entrance slit. The second gain is due to the fact that in conventional spectrometers each spectral interval is recorded in turn, while in Fourier spectrometers, the recording time of each spectral interval is equal to the recording time of the entire spectrum. Both factors together can give a gain in the value of the registered energy by four orders of magnitude.

A significant advantage of the method is also the absence of restrictions in the spectral resolution due to the size of the optical elements. Thus, the natural resolution limit for instruments using spatial dispersion is 0.02 cm^{-1}. At the same time, the serial industrial production of Fourier spectrometers with a resolution of up to 0.002 cm^{-1} has already been launched.

The third advantage of interference spectroscopy is the ease with which spurious light can be eliminated. This does not require any complex filter complexes, which are used in monochromators with diffraction gratings to remove unwanted diffraction orders.

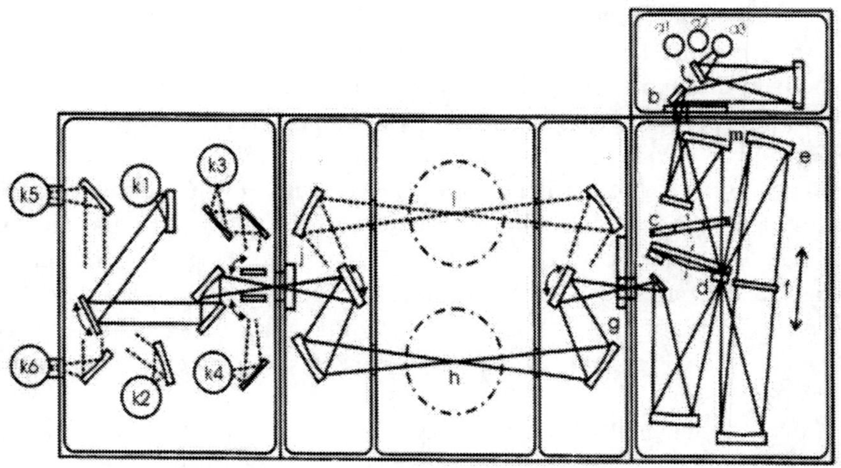

Figure 2. Fourier spectrometer Brucker IFS-113v: a3-source, b-entrance diaphragm, c-beam splitter, d-semitransparent mirror, e-reflecting mirrors, f-movable mirror, i, h- location of the samples, k1-Rollin detector.

A typical optical design of a Fourier spectrometer, for example, from Brooker IFS-113v, uses a Michelson interferometer (Figure 2). Passed through the entrance aperture (b), the light from the source (a3) falls on the collimator mirror (m) and by parallel light beam is directed to the beam splitter (c). In the low-frequency region, when the wavelength exceeds 25 μm, polymer films (here Mylar) are used as beam splitters. After the beam splitter, the transmitted and reflected beams fall on the reflecting mirrors (e), the quality and stability requirements of which in interferometers are very high: their surface should not deviate from the ideal by more than 1/20 of the wavelength corresponding to the short-wavelength limit of the device operation. The radiation emerging from the interferometer is focused by a mirror objective (n) in the place where the sample is placed if the absorption spectra are investigated. After that, the light is focused on the radiation receiver (k1). An important element of the optical scheme is the system for measuring the path difference between the mirrors of an interferometer using a He-Ne laser (not shown here). After conversion, its signal is used to create command pulses for reading readings from the radiation receiver in the receiving-amplifying system of the interferometer when the movable mirror is displaced. Thanks to such a system, the Fourier spectrometer becomes a device with a high accuracy in measuring the frequencies of spectral lines, and the accuracy is determined by the accuracy of determining the generation frequency of the reference laser.

The need for significant mathematical processing of the interferometer output data to obtain a spectrum made the computer (as already noted) an integral element of the device. At present, thanks to the fast Fourier transform algorithm, as well as the development of computer technology, calculations related to the Fourier transform can be performed on personal computers even for spectra containing hundreds of thousands of points.

Modern industrial Fourier-transform spectrometers such as "Brucker IFS-113v" are a powerful instrument for spectroscopic research. Two-thirds of the FIR spectra of polymers presented in this review were recorded with such devices. But even Fourier spectrometers, of course, they are not able to solve all the variety of problems in FIR spectroscopy. In some cases, difficulties arise due to the inevitable illumination of samples by broadband radiation from a thermal source for Fourier spectrometers, in others, due to low radiation power in the long-wavelength region, and in others, due to insufficient spectral resolution.

One of the alternative ways to solve the problem is a fundamentally new approach to the implementation of spectral studies in the submillimeter range, based on the successes of quantum electronics and nonlinear optics [11]. It is about the generation of submillimeter radiation during the propagation of a femtosecond laser pulse in a nonlinear medium. Due to the effect of optical detection, its wide spectrum is transferred from the optical range to zero frequency, where it covers the region up to 0.1 - 10 THz (3.33 - 333 cm^{-1}). The idea of a generator, as can be seen, is very simple, but its experimental implementation requires exceptional sophistication. In the last two decades, these difficulties have nevertheless been overcome thanks to the development of ultrashort pulse lasers. And now we can already say that femtosecond submillimeter spectroscopy, now called terahertz spectroscopy, is a serious competitor to the existing methods and, undoubtedly, it has a great future, primarily as a method of nonlinear spectroscopy at high levels of pulsed power. A recent Google search found 51,700 references for the term "terahertz spectroscopy" and 61,500 references for the term "long wavelength spectroscopy." The scientific literature now even speaks of a renaissance in low-frequency IR spectroscopy [6].

In addition to a pulsed femtosecond laser generating pulses with a duration of about 100 fs (1 fs = 10^{-15} s), terahertz spectrometers (in the

English-language literature THz TDS) include a terahertz (THz) radiation generator, which is a photoconductive (PC) antenna. The PC antenna consists of two metal electrodes located at some distance from each other on a semiconductor, usually gallium arsenide (GaAs) substrate. A voltage of the order of several kilovolts is applied to the electrodes. When the gap between the electrodes is illuminated by an ultrashort laser pulse, the concentration of charge carriers in the semiconductor sharply increases for a short time. The formed free carriers are accelerated by the field applied to the gap. as a result of which a short-term (of the order of units or tens of picoseconds) current pulse arises, which is the source of THz-radiation. THz radiation is then collected by parabolic mirrors onto the sample, and after passing through it, onto the detector.

The THz-radiation detector is designed and works, with the exception of some details, similar to the PC antenna. In this case, a current meter is connected to the electrodes instead of a voltage source. A current pulse is recorded, obtained by simultaneous illumination of the semiconductor with a terahertz and probe laser pulse. The current is proportional to the electric field of the terahertz pulse at the time of the arrival of the probe pulse. The terahertz field is recorded as a function of the time delay of the probe pulse. A terahertz pulse usually contains only a few field oscillations. By recording the waveform of a terahertz pulse after its interaction with the sample and then performing the Fourier transform from the waveform, it is possible to obtain the spectral characteristic of the sample in the range of frequencies present in the spectrum of the pulse. Due to the fact that the spectrum is obtained by recording the waveform (oscillogram) of the pulse, this method is called terahertz time-domain spectroscopy (in the English literature - Terahertz Time-Domain Spectroscopy, THz TDS) [13, 14].

A typical optical scheme of a terahertz spectrometer, for example, from THz-TDS 2004, Airspec, is shown in Figure 3.

The laser beam of a femtosecond sapphire laser with titanium with a generation wavelength of 780 nm and a pulse frequency of 100 fs at a power of 1200 mW is split into two beams using a splitter 2: a pump beam and a probe beam.

A more powerful pump beam is used to generate a terahertz pulse in the crystal of the generator 4, to which an electric field is applied, and the probe beam is used to create a temporary strobe on the photodetector 8, the signal from which reflects the profile of the electric field of THz radiation. A GaAs crystal is used as a photodetector, which is a photoconductive antenna that receives a THz signal. The signal from the photoconductive antenna of the detector, proportional to the electric field strength, then enters the signal registration and processing system 10-12.

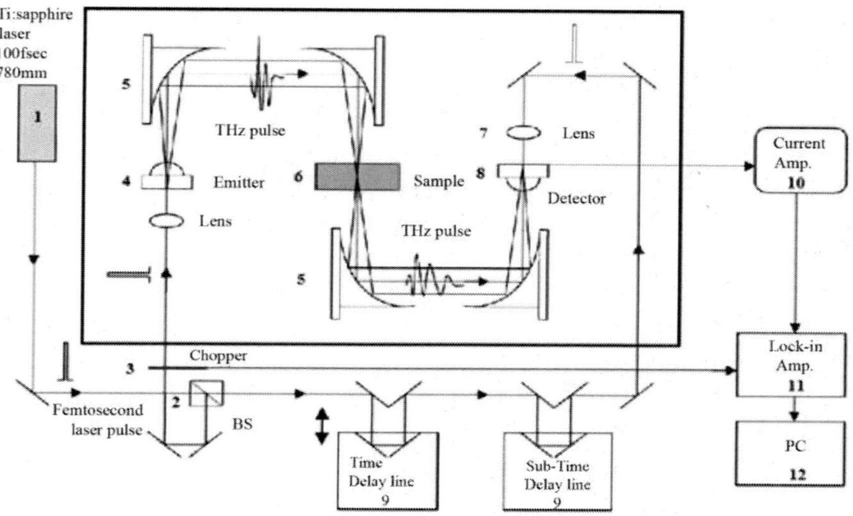

Figure 3. Terahertz spectrometer "THz-TDS 2004, Airspec". 1- femtosecond laser, 2- beam splitter, 3- chopper, 4- THz-radiation generator, 5- parabolic mirrors, 6- sample, 7- silicon lenses, 8- THz-radiation detector, 9- delay lines. 10- current amplifier, 11- synchronous amplifier, 12- personal computer.

The terahertz radiation emitted by the crystal of the generator often has a wide the directional diagram, therefore, parabolic mirrors 5 with a large numerical aperture are usually used to collect it. For effective detection of a terahertz beam, it must also be concentrated in a small region of space (with dimensions on the order of a wavelength), therefore, similar mirrors are used in the detector arm. The test samples 6 are placed in a terahertz beam between parabolic mirrors. Silicon lenses 7 are used to form the reference and probe beams. By changing the time delay between the THz pulse and the femtosecond laser pulse using the optical delay line 9, the signal from the THz detector is recorded, and the time dependence of the THz electric field amplitude is obtained. The spectrum of this signal is determined by Fourier transform, and the absorption spectrum is determined by subtracting two spectra recorded with the passage of the THz pulse through the sample and without it. To prevent the influence of absorption of THz radiation in water vapor in air, the optical scheme, where THz radiation passes, is placed in a sealed volume filled with dry nitrogen.

Comparing Fourier transform spectroscopy (FTS) operating in the low-frequency IR range with the method of terahertz spectroscopy, we note that the latter is distinguished by a number of features that give it significant advantages in spectral studies. First, THz-TDS uses a more stable and powerful radiation source - a laser pulse of hundreds of mW, duration 1-2 ps, which allows you to confidently and directly measure the amplitude and phase of a shorter terahertz pulse (FTS measures only the amplitude).

Second, since the absorption coefficient and refractive index obtained by the Fourier transform of the terahertz radiation transmitted through the sample are directly related to the amplitude and phase of this radiation, THz-TDS provides data in the form of a dielectric loss spectrum. (The role of the relaxation contribution in the formation of the low-frequency IR spectrum was discussed above). That is, calculate the complex dielectric constant $\hat{\varepsilon}(\omega)$ and the total refractive index $\hat{n}(\omega)$.

The latter is described in terms of the real part of the refractive index $n(\omega)$ and the imaginary part $ik(\omega)$, where $k(\omega)$ is the absorption index (sometimes also called the extinction coefficient): $\hat{n} = n(\omega) + ik(\omega)$. Using the speed of light in vacuum c and the circular frequency ω ($= 2\pi\nu$), the absorption coefficient $\alpha(\omega) = 2\omega k(\omega)/c$ can be obtained from the absorption index. Since the complex refractive index can be expressed in terms of dielectric losses: $\{\hat{n}(\omega)\}^2 = \hat{\varepsilon}(\omega) = \varepsilon'(\omega) - i\varepsilon''(\omega)$, it is possible to transform the spectrum of the absorption coefficient $\alpha(\omega)$, with which are usually dealt with in IR spectroscopy, into the spectrum of dielectric losses $\varepsilon''(\omega)$, using the following expressions: $\varepsilon'(\omega) = n(\omega)^2 - k(\omega)^2$ and $\varepsilon''(\omega) = 2n(\omega)\,k(\omega) = n(\omega)\,\alpha(\omega) \cdot c/\omega$. (This possibility is also essential in the terahertz IR region, since here resonant oscillatory processes coexist with relaxation ones).

Recall, in addition, that the absorption coefficient $\alpha(\omega)$, in turn, is related to the optical density (D) of the material: $D(\omega) = -\log_e(I/I_0) = \alpha(\omega)\,d$, where I is the intensity of the through the medium of light, and I_0 is the intensity of the incident light and d is the thickness of the material under study.

And third, perhaps the most important advantage of the terahertz technique is that it provides a 2-3 orders of magnitude higher signal-to-noise ratio than Fourier spectroscopy. This, in addition to allowing one to study strongly absorbing materials and operate in the wave number range from 0.9 to 90 cm^{-1} inaccessible to FTIR spectrometers, provides a high, about 20 GHz, frequency resolution. This solution "Brucker IFS-113v" can really provide bake only up to ~ 30 cm^{-1}.

If high resolution is not required, the high signal-to-noise ratio of the terahertz spectrometer recording system can be used to increase the spectrum recording speed.

Currently, all over the world, developments are underway to create methods for the use of THz radiation in the frequency range, especially for problems in medicine and biology. It is connected, formerly This is due, first of all, to the fact that in this range are the vibration frequencies of large groups of atoms that form a molecule and vibrations of hydrogen bonds of many organic substances of interest for biology and medicine (proteins, DNA molecules). They are very sensitive to the geometric shape of the molecule, its environment and play an important role in biochemical reactions. In addition, THz radiation is not ionizing and, therefore, hazardous to biological objects, like the commonly used X-ray, which makes it possible to use it in vivo. All these factors make it possible to use THz spectrometers for many studies: identification of biomolecules, including the determination of their mutations and various conformational states; study of biological tissues, in particular, subsurface layers, which is directly related to the diagnosis of the depth of their damage (for example, due to a burn), the presence of tumors, necrosis and other pathological processes.

The third way is the use of synchrotron radiation to obtain low-frequency IR spectra of polymers. Just like a heated black body, electrons in a synchrotron accelerator emit electromagnetic radiation in a very wide spectral range - from radio frequency to X-ray. If we talk about the long-wavelength region, the main advantage of synchrotron radiation is that its intensity falls off as the first power of frequency (and not as the fourth for a black body). Such a source is a dream for Fourier spectroscopists, but again everything rests on technical "details." Accelerators providing acceptable radiation powers are huge energy-intensive installations, which cost several orders of magnitude higher than the cost of the Fourier spectrometers themselves. For this reason, the question of using for spectroscopic purposes, synchrotron radiation, apparently, can be set only in the sense of the implementation of "waste-free technologies" on the available accelerators.

References

[1] Yaroslavl NG, Zheludov BA, Stanevich AE (1956) *Opt. and spectrum.* 1: 507.

[2] Martin DH (1963) *Contemp. Phys.* 4: 187.

[3] Hadni A (1963) *Spectrochemica Acta!* 9: 793.

[4] Naumenko VM, Fomin VI, Eremenko VV (1967) *PTE* 5: 223.
[5] Richards PL (1964) *JOSA* 54: 1474.
[6] Yaroslavl NG, Stanevich AE (1958) *Opt. and spectrum.* 5: 384.
[7] Robinson DW. (1959) *JOSA,* 49: 966.
[8] Russel JW, Strauss HL (1965) *Appl. Opt.,* 4: 1131.
[9] Ryzhov VA, Tonkov MV (1973) *Molecular Spectroscopy, collection of articles.* 2: 108 Ed. Leningrad State University, Leningrad.
[10] Bell RD (1975) *Introduction to Fourier Spectroscopy.* Ed. "Mir", Moscow.
[11] Grischkowsky G (1988) *Proc. 4 Intern. Conf. Infrared Phys*. P. 51, Zurich.
[12] Mantsch H, Naumann D (2010) *J. Molec. Struct.* 964: 1.
[13] Dexheimer (2007) (Ed.), *Terahertz Spectroscopy: Principles and Applications.* CRCPress, New York.
[14] Schmuttenmache CA (2004) *Chem. Rev.* 104: 1759-1779.

Author's Contact Information

Valery A. Ryzhov
Researcher
Ioffe Institute
Russian Academy of Sciences
St. Petersburg, Russia
v.ryzhov@mail.ioffe.ru

Index

E

F

G

H

I

L

M

O

P

R

S

T

V